UNDERSTANDING DIGITAL ELECTRONICS

UNDERSTANDING DIGITAL ELECTRONICS

by

R. H. WARRING

LUTTERWORTH PRESS · GUILDFORD, SURREY

First published 1982

ISBN 0 7188 2521 7

Photoset and printed photolitho in Great Britain by
Ebenezer Baylis and Son Limited,
The Trinity Press, Worcester, and London.

Contents

CHAPTER 1

Getting to Grips With Digital Electronics

Most people are familiar with 'ordinary' electrical circuits where voltages and currents can be measured with dial-type meters, or *analogue* meters as they are technically described. Such meters also follow changes in voltage (or current) in a smooth *stepless* manner. That is what analogue really means in electronics: signals or physical quantities which can vary in a smooth-changing, stepless manner.

Everyone knows that certain analogue quantities, e.g. clock time, can be presented in the form of a *digital display* as an alternative to analogue display (a clock with hands). Apart from the difference in presenting time — visible numbers instead of position of the hands — one difference is immediately apparent. The digital display changes in jumps, from one number to the following number. Clock or watch hands move *progressively* from one position to another. (Actually they move in tiny jumps when driven by a gear train, but they are still analogue rather than digital movements. A better example of a *pure* analogue movement is a meter reading voltage or current with no gear train involved. Again most meters can be constructed to provide a digital display of the quantity being measured, which changes a digit at a time with no 'intermediate' values).

So 'digital' really means working in steps. Converting varying analogue quantities into digital read-outs is only a small — but important — part of digital electronics. Its principle attraction is that it is very positive and accurate, working a step at a time. It is more *positive,* for instance, to *count* on the fingers that four (fingers) plus five (fingers) totals nine (fingers), than remembering that four plus five equals nine.

The same argument applies even more so to much more complex arithmetical problems. Counting, a step at a time, should *positively* give the right answer. Using digital electronic devices, some of which are capable of step-by-step counting at a rate of up to 1000 *million* times a second, can also be an extremely rapid method of working.

But counting, and performing arithmetical and other mathematical functions, is only one of the capabilities of digital devices. Another important application is in performing *logic* functions. Logic, boiled down to its basic form is really a question of 'yes' or 'no' (or 'true' or 'false'). A whole host of circuit design problems — from simple switching circuits to complete microprocessors and computers — can be implemented in hardware form by basic logic elements on a step-by-step (digital) basis of working, again at fantastic speeds.

Design problems in logic can be written out in three different ways: in 'plain' language (block logic); possible permutations and what they give (truth tables); or mathematically (using Boolean algebra). Solutions can then be implemented as working circuits employing *logic devices* suitably interconnected. These various subjects represent whole new fields to study and become familiar with, especially for anyone with little or no previous knowledge of digital electronics — and there is always more to learn.

This book attempts to break down digital electronics into separate subjects which can be read and studied separately, if necessary. Start with Logic Circuit Devices (Chapter 5) if you like. Or, if you are 'mechanically' minded, start with Symbols and Switches (Chapter 3) for a clear understanding of how electronic logic devices are nothing more than combinations of on-off switches. You can always refer to earlier chapters to fill-in on binary arithmetic and mathematical logic, etc.

You will need to study, or at least read through, the earlier chapters for a proper appreciation of the various application circuits contained in Chapter 8 onwards. With that proviso much of the mystique of digital electronics should become clear — except probably for Chapter 13 and Appendix III. These are there to explain some of the more complex techniques related to modern digital computers. Stop before there, if you like. After all, computer programming is a subject which even the experts struggle with!

CHAPTER 2

Binary Arithmetic and Truth Tables

Digital systems employ devices which function in only two states, hence they are known as *binary* devices and operate in a binary manner (binary simply means two). These two states may be described in various ways, for example:

on-off — or closed-open for switching devices

1-0 — for counting or computing devices

true-false — or yes-no for logic devices

pulse-no pulse — for trigger circuits

HIGH-LOW — for practical digital circuits where voltage levels are relative (i.e. a low signal is not necessarily zero).

All mean the same thing in function — one state *or* the other, with no intermediate state. This is the whole basis of digital (binary) working.

Corresponding binary numbers are based on just two digits, 1 and 0. Individual digits in a binary number then represent equivalent powers of 2, instead of powers of 10 as in the decimal system. A particular advantage of the binary system is that there are no 'multiplication tables' as such and any problem involving addition, subtraction, multiplication, division, etc., can be broken down into a series of individual binary operations, with each switching element in the system being continuously used (i.e. either in the 'on' or 'off' state).

Compared with the decimal system, binary numbers are tedious as a written language. Here, for example, are the binary equivalents of decimal numbers from 1 to 32.

decimal	binary number	decimal	binary number
1 (2^0)	1	17	10001
2 (2^1)	10	18	10010
3	11	19	10011
4 (2^2)	100	20	10100
5	101	21	10101
6	110	22	10110
7	111	23	10111
8 (2^3)	1000	24	11000
9	1001	25	11001
10	1010	26	11010
11	1011	27	11011
12	1100	28	11100
13	1101	29	11101
14	1110	30	11110
15	1111	31	11111
16 (2^4)	10000	32 (2^5)	100000
		and so on	

Remembering that the binary system is based on powers of 2, the simplest way to derive the binary number equivalent of a large decimal number is to subtract the highest power of 2 contained by the number; then subtract the highest power of 2 from the remainder; and so on until only a 1 or 0 is left as a remainder. For example, suppose we want the binary number for the decimal number 269:

The highest power of 2 within this number is $2^8 = 256$

This leaves $269 - 256 = 13$

The highest power of 2 within this remainder is $2^3 = 8$

This leaves $13 - 8 = 5$

The highest power of 2 within this remainder is $2^2 = 4$

This leaves $5 - 4 = 1$.

The corresponding number is thus $2^8 + 2^3 + 2^2$ with remainder 1, i.e.

$$
\begin{array}{rrll}
2^8 = & 100000000 & = 256 & \text{decimal} \\
2^3 = & 1000 & = 8 & \\
2^2 = & 100 & = 4 & \\
\text{remainder} = & \underline{1} & = \underline{1} & \\
& 100001101 & = 269 & \text{decimal}
\end{array}
$$

(The column on the right is written out as a check).

The binary number is long: 9 digits or *bits* as they are called. Nevertheless it 'counts' in a system involving only 1 or 0, so it can readily be handled by digital devices. Nor does the number of bits to be handled in a calculation represent any practical limitation. The speed at which these devices can work is extremely high. Here, for example, is the number of bits (digits) different types of digital circuit devices can handle *per second:*

MOS (metal-oxide-semiconductor) 3-4 *million*
CMOS (complementary-metal-oxide semiconductor) 10-15 *million*
HTL (high-threshold-logic) 20 *million*
DTL (diode-transistor-logic) 35 *million*
RTL (resistor-transistor-logic) 80 *million*
TTL (transistor-transistor-logic) 170 *million*
ECL (emitter-coupled-logic) 250-1000 *million*

To simplify matters with large numbers a hybrid system known as a *binary coded decimal* is normally used. Here separate groups of binary digits are used to express units, tens, hundreds, etc. Since each binary group needs to be able to accommodate a count of up to 9, it must consist of four digits, i.e. to accommodate 9 it must run to 1001 (see Table on p. 10).

The number 269 (normally 100001101 in the binary system) would thus, as a binary coded decimal, become:

0010 : 0110 : 1001
Equivalent to   2   6   9   in decimal numbers

For the next number up (270), the right-hand binary group would change to 1010, representing 10 but immediately carry this forward into the next group. The binary coded decimal would then read:

0010 : 0111 : 0000
Equivalent to   2   7   0

Binary coded decimal systems are described in detail in Chapter 7.

Truth Tables

Truth tables are an easily understood way of representing the working of digital devices. They are widely used in solving circuit design problems, together with Boolean algebra (*see* Chapter 4). A truth table simple lays out the complete range of signal states for a device in terms of 1 (signal on) or 0 (signal absent).

Starting with the simplest device, a NOT gate, there is one signal input A (which may have a state of 0 or 1); and one signal output S. Since a NOT gate gives inversion, the state of output S will be the invert or *opposite* of input A. The truth table then reads like this:

A	S
0	1
1	0

This fully expresses all the possible working states (two in this case) of a NOT logic element.

All other devices have more than one input, a basic rule to follow here in compiling a truth table is that with *series logic,* all the inputs must be 1 before the output can become 1; and with *parallel logic* the output will always be 1 if *any* of the inputs is also 1. Thus, the basic truth tables for such devices, written for two inputs are:

Series logic				Parallel logic		
A	B	S		A	B	S
0	0	0		0	0	0
1	0	0		1	0	1
0	1	0		0	1	1
1	1	1		1	1	1

OR *Truth Table*

The truth tables for an OR logic element with two inputs is:

A	B	S
0	0	0
1	0	1
0	1	1
1	1	1

It is, in fact, an example of a *parallel logic* device. Expanded to cover more than two inputs, the same basic rule applies: S will equal 1 when any input equals 1. Thus for a four input OR logic element:

A	B	C	D	S
0	0	0	0	0
1	0	0	0	1
0	1	0	0	1
0	0	1	0	1
0	0	0	1	1
1	1	0	0	1
1	0	1	0	1
1	0	0	1	1
1	1	1	0	1
1	1	0	1	1
1	0	1	1	1
0	1	1	1	1
0	1	0	1	1
0	1	1	1	1
0	0	1	1	1
1	1	1	1	1

There are, indeed, sixteen different states possible with any four input gates; and in the case of the OR device, fifteen of these will give an output signal. This is equally well explained by 'mechanical' thinking since parallel logic is equivalent to a number of on-off switches connected in *parallel*. Any *one* switch which is 'on' will pass a signal.

AND *Truth Table*

The AND gate is series logic, so the basic truth table with two inputs is:

A	B	S
0	0	0
1	0	0
0	1	0
1	1	1

Written out for four inputs this becomes:

A	B	C	D	S
0	0	0	0	0
1	0	0	0	0
0	1	0	0	0
0	0	1	0	0
0	0	0	1	0
1	1	0	0	0
1	0	1	0	0
1	0	0	1	0
1	1	1	0	0
1	1	0	1	0
1	0	1	1	0
0	1	0	1	0
0	1	1	1	0
0	0	1	1	0
1	1	1	1	1

Again there are sixteen possible different states, but only one provides an output S = 1. With sixteen 'switches' connected in *series* the path through them from input to output remains broken until *all* the switches are 'on'.

The immediate reaction to these two examples is probably a feeling that it is much simpler to work in terms of 'switching' equivalents than truth tables. And for very simple problems in digital logic it is. However, most problems require a combination of logic devices to provide the solution, which may involve both series and parallel logic. Drawing out the switching circuits can then become a more elaborate process than plotting truth tables – and be more prone to making mistakes in the process.

Truth tables for other logic elements are given below, written for two inputs. They can be expanded to present truth tables for more than two inputs by following the same established *pattern* of behaviour.

Truth Table for NOR logic:

A	B	S
0	0	1
1	0	0
0	1	0
1	1	0

This can be identified as 'inverted' series logic, when the extension to four inputs would follow as:

A	B	C	D	S
0	0	0	0	1
1	0	0	0	0
0	1	0	0	0
0	0	1	0	0
0	0	0	1	0

and so on.

Note also that *inversion* has changed the *parallel* logic of an OR to *series* logic in the case of NOR (NOT-OR). The significance of this occurs frequently when working with Boolean algebra (*see* Chapter 4).

Truth Table for NAND *logic:*

A	B	S
0	0	1
1	0	1
0	1	1
1	1	0

This is 'inverted' parallel logic. Inversion has changed the series logic of AND to parallel logic in the case of NAND (NOT-AND).

Combinations of Logic Elements

The state of combinations of logic elements can be expressed in the same way as a truth table. Suppose, for example the design requirement is to provide for input signals A or B to produce an output signal *only* in combination with a third input signal C. (For example, A and

B are trainee operators who can only give a command signal to a machine when the instructor (C) also adds his own signal.)

This can be provided by OR and AND logic:

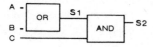

Writing the truth table for the OR device first and calling the output S1:

A	B	S1
0	0	0
1	0	1
0	1	1
1	1	1

S1 is now one of the inputs to the AND device. The truth table for this device is then:

S1	C	S2
0	0	0
1	0	0
0	1	0
1	1	1

The *combined* truth table can then be written as:

A	B	C	S2
0	0	0	0
1	0	0	0
0	1	1	1
1	1	0	0
1	1	1	1
1	0	1	1

Note. Number of 'states' is equal to the number of devices multiplied by the number of inputs to each device, i.e. $3 \times 2 = 6$ possible states in this case.

Plotting Truth Tables First

A truth table can be drawn up as a starting point in design. For example, suppose the problem is concerned with a control circuit to start and operate a machine under the following conditions:

A. Signal from operator standing by machine

OR D. Signal from a remote start position

AND B. Signal confirming guard is in place
AND C. Signal from detector showing workpiece is in place
The machine must *not* start under any other conditions.

There are four inputs to consider, which will result in sixteen possible combinations or states. This establishes the basis for writing out a
five-column, sixteen-line truth table. On the output column, 1 must
only appear when A = 1 *or* D = 1 *and* B = 1 *and* C = 1. All the other
combinations of A, B, C and D must give S = 0.

A	B	C	D	S
0	0	0	0	0
0	0	0	1	0
0	0	1	0	0
0	0	1	1	0
0	1	0	0	0
0	1	0	1	0
0	1	1	1	1
1	0	0	0	0
1	0	0	1	0
1	0	1	0	0
1	0	1	1	0
1	1	0	0	0
1	1	0	1	0
1	1	1	0	1
1	1	1	1	1

Fig. 2.1 shows this truth table implemented with logic devices and
also with mechanical switches.

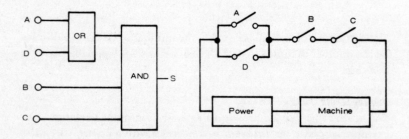

Fig. 2.1. Logic solution (left), and implementation with mechanical switches.

CHAPTER 3

Symbols and Switches

One of the most confusing things about the use of symbols representing the various logic elements or gates is that the original (and literally 'logical') way of designating them in the form of annotated blocks has largely been abandoned in favour of representative symbols, the significance of which is not apparent until you are familiar with them. Even then misunderstanding can arise since over the years different symbols have been used to illustrate the same function(s). Various attempts have been made to standardize symbols, CETOP standard recommendations being the ones most widely adopted in Europe. American literature remains tied to US MIL standard symbols.

Another source of confusion is that different letter symbols may be used to designate input(s) and output(s), particularly for the basic devices, e.g. A, B, C, or XYZ, for inputs; S, Q or Z for output. This is not particularly important if the application is clear, but can cause confusion with more complex devices where specific symbols (and sometimes different symbols) are used to designate specific IC (integrated circuit) terminals, e.g. Ck (or ϕ) for *clock* input, D for *data* input, etc.

Having probably confused the issue even more by these paragraphs, the best thing to do is to go back to the beginning and start again!

The simple *block* method of symbolizing logic elements is obvious, readily readable, and really needs no further description. All symbols are in the form of a rectangular block and the function written inside. *Input* lines are added to the left side of the block; and an output line to the right. Thus Fig. 3.1 shows a number of representative logic elements with two or more inputs and one output each. (The NOT of course only has one input). For the sake of consistency, separate inputs are designated A, B, C, etc. The output line is designated S (S for signal out). The value 1 is used to designate a signal present, and an 0 represents no signal at that line.

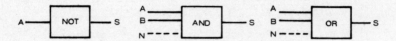

Fig. 3.1. Examples of block logic symbols and annotation.

Such block symbols are now seldom used, except in elementary textbooks. It is thus necessary to know the alternative forms of other basic symbols and familiarize yourself with them. We will deal with each function separately and also illustrate this with its corresponding *mechanical* switching function, restricting devices to two inputs for simplicity. The switching function is shown as 1 in the 'operated' position, and 0 in the 'off' position.

Order of presentation in each case, from left to right, is block form, mechanical switch equivalent, CETOP, US MIL, British Standard.

YES

This is a 1-input, 1-output device, output and input always being the same, i.e. A = 0, S = 0, or A = 1, S = 1, where A is the input signal and S is the output signal.

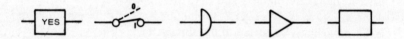

Fig. 3.2. YES logic symbols.

NOT (INVERTOR)

This 1-input, 1-output device works the other way round to YES. If there is an input at A, there is no output at B; and vice versa (i.e. A = 1, S = 0; or A = 0, S = 1). The symbols show this inverted mode of working by means of a circle ● or 0 on the output side.

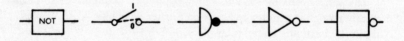

Fig. 3.3. NOT logic symbols.

AND

Here the circuit through the device is completed only when input A = 1 *and* B = 1, when S = 1. In mechanical form, two switches in series.

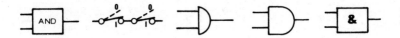

Fig. 3.4. AND logic symbols.

NAND

The inverted form of AND, where there is an output in all states of input *except* when A = 1 *and* B = 1 (when S = 0). The symbol is the same as for AND with the inversion mark added.

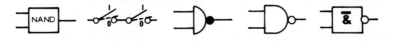

Fig. 3.5. NAND logic symbols.

OR

This is the equivalent of a parallel switching circuit. When either switch is closed (or both switches are closed), there is an output.

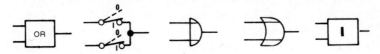

Fig. 3.6. OR logic symbols.

NOR

The inverted form of OR, so once again the symbols have the inversion mark added. Note from the switch circuit that the switches are normally closed and there is an output (S = 1) only when both A = 0 and B = 0.

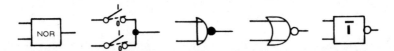

Fig. 3.7. NOR logic symbols.

EXCLUSIVE OR

This is a special form of AND logic providing an output only when one particular input is equal to 1. If a 1 appears at the other input it is inhibited or inverted to 0. Basically, in fact, it is an AND gate with one input inverted, as the symbols show. (The A input being inhibited in these diagrams, but it could equally well be B.)

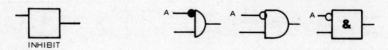

Fig. 3.8. INHIBIT symbols.

MEMORY

Memory function is performed by a flip-flop (FF) and here we have a 'store' state rather than a switching function. The output state depends on the last input applied and is maintained when the inputs are resumed. Basic symbols are as shown in Fig. 3.9.

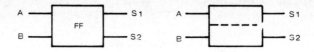

Fig. 3.9. Basic symbols for a memory unit (flip-flop).

In practice there are different *types* of flip-flop, each of which is given its specific symbol and inputs and designated accordingly, e.g. S and R for a S – R flip-flop; J and K for a J – K flip-flop; D for a D-type flip-flop; and T for a T-type flip-flop. Outputs are then designed Q and $\bar{Q}$. In addition the flip-flop may have a *clock* signal input (designated C, Ck, or ϕ); a *clear* signal input (Cr), and a *preset* input (Pr), depending on type. These symbols are illustrated in Fig. 3.10.

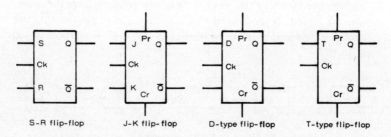

Fig. 3.10. Symbols for different types of flip-flops.

For more detailed information on flip-flops, *see* Chapter 6.

Simple Switching Functions

As an example of the application of logic to the design of switching circuits using digital devices, take the problem of designing a circuit for switching a single light on and off from two separate points — a common arrangement in the hallway or stairway of houses.

The basic requirements are that there are two possible inputs (switches) — call them A and B — which may be either on or off. When either A *or* B is on, there is an output (i.e. circuit completed to light the bulb). Neither A nor B can be 'on' at the same time. If one is 'on', operating the other switch will switch the light 'off'.

Written in the form of a truth table:

$$1 = \text{switch 'on'} \begin{cases} \begin{array}{ccc} A & B & L \quad (1 = \text{light on}) \\ 0 & 0 & 0 \\ 0 & 1 & 1 \\ 1 & 0 & 1 \\ 1 & 1 & 0 \end{array} \end{cases}$$

Expressed in equation form (*see* Chapter 4 for explanation):

$$L = A\bar{B} + \bar{A}B$$

i.e. A 'on', B 'off' *or* A 'off', B 'on' will light the lamp.

A further equation can now be written expressing the combinations that *do not* produce an output (i.e. do not switch the light on) call this D (dark):

$$D = \bar{L} = AB + \bar{A}\bar{B}$$

Applying de Morgan's theorem:

$$L = \overline{(AB + \bar{A}\bar{B})}$$
$$= (A + B)(\bar{A} + \bar{B})$$
$$= A\bar{A} + A\bar{B} + \bar{A}B + B\bar{B}$$
$$= A\bar{B} + \bar{A}B$$

This merely restates, and proves the validity of, the first formula. However, it also provides a second equation for implementing the requirements specifically in binary (on-off) elements required to produce the desired switching circuit.

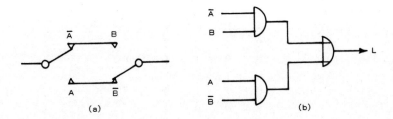

Fig. 3.11. First solution to switching problem.

Taking the first equation $L = \overline{A}B + A\overline{B}$, Fig. 3.11a shows this implemented in terms of mechanical switches (or relay contacts); and Fig. 3.11b shows the answer implemented in terms of logic gates.

Taking the second equation $L = (A + B).(\overline{A} + \overline{B})$, Fig. 3.12a shows this implemented in terms of mechanical switches (or relays); and Fig. 3.12b shows this circuit implemented in terms of logic gates.

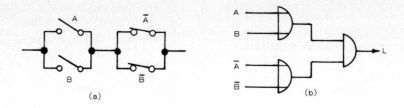

(a)   (b)

Fig. 3.12. Alternative solution to switching problem.

Fig. 3.11a is obviously the best practical solution, since it involves only half the contacts (series logic as opposed to parallel logic). In the case of the gate solutions the choice is not so obvious. It largely depends on the type of gates most readily available. Fig. 3.11b requires two AND gates and one OR gate. Fig. 3.12b requires two OR gates and one AND gate.

Treatment of these solutions may seem overcomplicated for the problem involved. Basically they are presented to show the principle of digital switching circuit design with a simple, easily understood example. Suppose we take it one step further to derive suitable circuitry for switching a light on from any of the *three* different switch points?

Starting point is to draw up the truth table:

A	B	C	L
0	0	0	0
0	0	1	1
0	1	0	1
0	1	1	0
1	0	0	1
1	0	1	0
1	1	0	0
1	1	1	1

This establishes all the possible input conditions, but does not give any immediate clue as to possible circuit design without drawing out each combination in detail.

Deriving a formula from the truth table (or original 'logic' requirements):

$$L = A\bar{B}\bar{C} + \bar{A}B\bar{C} + \bar{A}\bar{B}C + ABC$$

This will factorize as follows:

$$L = \bar{C}(A\bar{B} + \bar{A}B) + C(\bar{A}\bar{B} + AB)$$

It is now possible to simplify to some extent by calling $A\bar{B} + \bar{A}B = X$. Then, since $A\bar{B} + \bar{A}B = \overline{\bar{A}B + \bar{A}\bar{B}}$

$$L = \bar{C}X + C\bar{X}$$

Solutions to this equation implemented in the form of both mechanical switches and exclusive OR logic gates are given in Fig. 3.13.

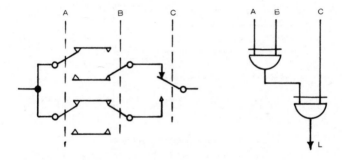

Fig. 3.13. Three-position light switching solution.

Series and Parallel Working

As already explained, digital devices may operate in *series* or *parallel* modes. The difference is simple to explain diagramatically.

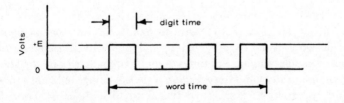

Fig. 3.14. Series working with positive logic (1 = + E).

With *series* operation, binary digits are expressed by voltage levels in a single output wire displaced in *time*. Thus a complete signal representing the binary number, say, 100101 would be as shown in Fig. 3.14. Note this is shown for positive working. It could equally well be given by *negative* working. In this case the 0 level could be + E, with each pulse appearing as an 0 value; or alternatively the 0 level being an 0 value and each pulse level being – E.

With *parallel* operation each digit is allocated a separate line. Outputs then appear simultaneously on each line — Fig. 3.15.

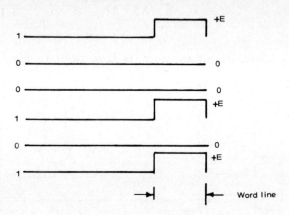

Fig. 3.15. Parallel working with positive logic (1 = + E).

Again these could appear as a negative instead of a positive voltage value representing a '1'. In many practical circuits, too, the change in voltage or signal *swing* may be from some nominal voltage representing condition '0' to some more positive (or more negative) voltage representing a '1'. In such cases the description HIGH or H is commonly used to designate a '1' signal and LOW or L a '0' signal. In other words, HIGH (or H) is used instead of '1'; and LOW (or L) instead of '0'.

Series working may appear the logical choice since it needs only one digital device or output wire to handle any number of digits. Parallel working has the disadvantage of requiring *n* devices or output wires to handle *n* digits, or *n* times as much circuit hardware to handle the same information. However, it has the advantage of being *n* times as *fast* as series working. In practical circuits, however, both the number of components used and operating time may be modified by other factors.

SIMPLE ELECTRONIC SWITCHES

A bipolar junction transistor can readily work as a switch although its characteristics are not ideal for this purpose. The most usual way of working is in the *saturation mode,* when the transistor has two stable states — one passing no current (except for leakage current) corresponding to 'off', and the other in the saturated state passing maximum current and corresponding to 'on' — Fig. 3.16.

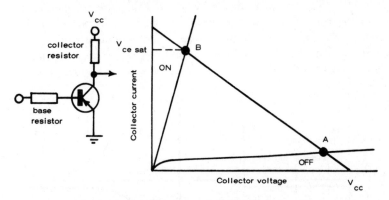

Fig. 3.16. Switching characteristic of a bipolar transistor.

In the 'off' condition the collector voltage approaches Vcc. In the 'on' condition the collector voltage is Vce, which is typically of the order of 0.15 to 0.6 volts. However, the transistor is now capable of passing a (relatively) large voltage.

Simplified design parameters for such a switching circuit are:

Base current $I_B = Vcc/R_B$

Collector current $I_C = Vcc/R_C$

$I_C = h_{FR} \times I_B$

Bias resistor $R_B = h_{fe} \times R_C$

where h_{fe} is the current gain of the transistor in the saturated mode. R_C is the load resistance in the collector line.

All these formulas are approximate only. In practice it is usually necessary to make the value of R_B a fraction of the theoretical value (e.g. ¼) to allow for tolerances and ensure that the transistor remains saturated over a range of input voltages.

FET's can also be used in a similar manner as switches. They do not suffer from the same propagation delay present with bipolar transistors, but still have *turn-on* and *turn-off* delays due to interelectrode capacitance. These are of a similar order to, or higher, than the delay times characteristic of bipolar junction transistors.

Improving Transistor Switch-off Times

A direct method of reducing the switch-off time of a transistor is to reverse-bias the base, but any such bias must not be allowed to exceed the reverse voltage limit of the transistor otherwise it will be damaged. An alternative method is to *clamp* the base voltage to prevent the transistor from becoming saturated during the switch-on period. This, too, has its limitations so when fast switching times are required from bipolar transistors *current switching circuits* are normally employed in which the transistor never becomes saturated nor cut-off.

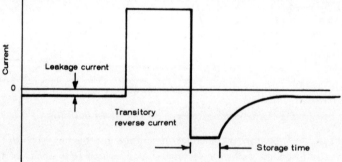

Fig. 3.17. Switching characteristics of a diode.

Diode Switches

Diode switching characteristics are illustrated in simplified form in Fig. 3.17. When reverse biased there is only a very small leakage current. Application of forward voltage results in an immediate step to + V (forward conduction). The next application of reversed (– V) voltage, however, produces a transient due to *stored charge effect,* which then decays to the leakage current value. The peak transient reverse current can approach – V/R as a maximum, where R is the resistance in the circuit. The time to reverse this charge, or *storage time,* varies with the type of diode and construction. In the case of ordinary diodes it can be a matter of milliseconds, reducing to nanoseconds in the case of high-speed switching diodes.

Schottky

The *Schottky diode* differs from conventional diodes in having a metal-to-semiconductor function at which rectification occurs. It has specific advantages over conventional junction diodes in that it does not exhibit carrier charge storage effects, thus enabling much faster switching speeds to be achieved. The voltage drop across a Schottky diode is also much less than that of an ordinary diode for the same forward current.

Diodes are commonly used as a clamp between the base and emitter of a transistor to prevent the transistor from entering saturation and to minimize propagation-delay time. It is readily possible to combine a Schottky clamping diode with a transistor as an integral device. Such a combination is called a *Schottky transistor.*

Unijunction Transistors

Unijunction transistors have two base contacts and an emitter. They become conductive (switch on) at a particular *firing* voltage, which typically ranges from 0.5 to 0.85 of the supply voltage. A particular application of unijunction transistors as switching devices is to generate short pulses when supplied with a varying supply voltage, when pulse rates of up to 1 MHz are readily obtainable.

Thyristors and TRIACs (SCRs)

A *thyristor* is basically a silicon diode with an additional cathode electrode known as a *gate.* If the gate is biased to the same potential as the cathode, it will not conduct in either direction (except for a small leakage current). However, if the gate is biased to be more positive than the cathode, the thyristor will behave as a normal diode, i.e. work as a switching element triggered by the application of a positive pulse to the gate.

The TRIAC is similar in construction except that it has both a cathode and anode gate, hence it can be triggered by both positive and negative pulses. Both are also known as *silicon controlled rectifiers* (SCRs). They are essentially *alternating current* switches — a thyristor being triggered by the positive half of an AC voltage and a triac by both positive and negative halves of an AC voltage. Typical basic switching circuits are shown in Fig. 3.18.

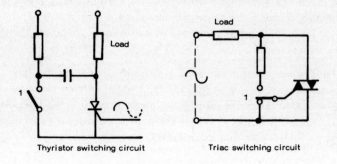

Thyristor switching circuit   Triac switching circuit

Fig. 3.18. Thyristor and Triac AC switches.

'Bounce-free' Switches

Mechanical switches commonly suffer from contact 'bounce' when closed, which can give a spurious signal (especially when switching at rapid rates). This can be avoided by employing a 'bounce-free' (or 'no bounce') switch. An example is shown in Fig. 3.19, employing an S – R flip-flop as a follower for a mechanical switch. The effect of any contact bounce is now to raise both inputs to the flip-flop to logic 1, leaving the outputs unaffected.

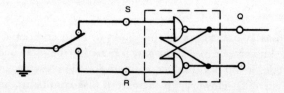

Fig. 3.19. 'Bounce-free' switch.

CHAPTER 4

'Mathematical Logic (Boolean Algebra)

Logic functions can be expressed by symbols, truth tables, or mathematically, the latter being known as Boolean algebra. It is named after George Boole, who devised the system of representing logic through a series of algebraic equations as long ago as the middle of the last century. Until the appearance of the first electronic computers (in 1938) Boolean algebra was regarded as an academic mathematical exercise. Today it is a tool widely used by designers of logic circuits.

The basic symbols used in Boolean algebra are:
● meaning a *series* condition or AND logic
+ meaning a *parallel* condition or OR logic
− meaning negation or opposite condition or NOT logic.

At this stage it is best to 'forget' conventional arithmetic where . means multiply and + means add, otherwise Boolean algebra will seem confusing at first. But multiplication and addition *do* enter into working with Boolean equations, as will be explained later.

Basic logic symbols are again shown in Fig. 4.1 with equivalent equations in Boolean algebra.

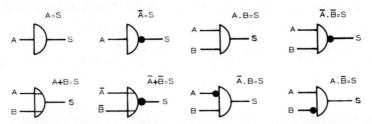

Fig. 4.1. Examples of annotated logic symbols with equivalent Boolean equations.

YES logic represents a simple continuous condition, i.e. the output(s) will be the same as the input(A). The corresponding mathematical equation is obviously $A = S$.

NOT logic represents a negation or opposite condition between input and output. Here the negation sign is used in the mathematical equation which becomes $\bar{A} = S$ (or $A = \bar{S}$).

AND logic requires that inputs A *and* B are both present before there is any output (a series condition), so the mathematical equation becomes A.B = S.

NAND logic is the negation of AND, so here the equation becomes $\overline{A.B}$ = S. Alternatively this equation can take the form A.B = $\overline{S}$, which implies the same logic.

OR logic represents a parallel condition in that an input must be present at either A *or* B before there is an output (a parallel condition). In this case the + sign applies and the mathematical equation becomes A + B = S.

NOR logic is the negation of OR, so the negation sign is added, i.e. $\overline{A} + \overline{B}$ = S.

The above basic equations are given for just two inputs. Exactly the same forms apply where there are more inputs. For example, the equation for an AND gate with five inputs would be:

$$A.B.C.D.E = S$$

With the exception of NOT (which has only a single input and can only invert signals), each of the expressions for a logic function can be rearranged to obtain the others. This is a useful tool when designing logic circuits for it enables the required functions to be rendered in the *same* logic dependent on the availability, or preferences, for particular components, e.g. all in OR logic; all in AND logic; or all in NAND logic. This is done largely by using a NOT function (or single input NOR gate) as an inverter where necessary and the principles established by de Morgan's theorem. Basically this states that inversion changes the state of the logic each time it is applied, i.e. from ● to + , or + to ●

For example, starting with the AND function:
$$A.B = S \text{ (Fig. 4.2a)}$$
Inversion will change the AND (.) to OR (+) logic:
$$\overline{A} + \overline{B} = \overline{S} \text{ (Fig. 4.2b)}$$

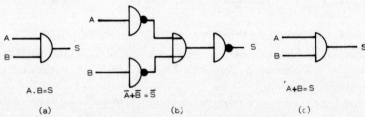

A.B=S $\overline{A}+\overline{B} = \overline{S}$ A+B=S

(a)   (b)   (c)

Fig. 4.2. AND function from NOT and OR devices.

Inverting again will give a positive output:

$$\overline{\overline{A} + \overline{B}} = \overline{\overline{S}}$$

which is the same as:

$$A + B = S \text{ (Fig. 4.2c)}$$

In other words double inversion has changed the function of an AND gate into OR logic working.

At this stage the basic rule to remember is that inversion changes the *sign* of the equation (except in a NOT gate) as well as changing the input, i.e.

Logic	equation for positive output	with inversion
OR	$A + B = S$	$\overline{A}.\overline{B} = \overline{S}$
NOR	$\overline{A}.\overline{B} = S$	$A + B = \overline{S}$
AND	$A.B = S$	$\overline{A} + \overline{B} = \overline{S}$
NAND	$\overline{A} + \overline{B} = S$	$A.B. = \overline{S}$

OR logic

Working with OR logic throughout, equations must be worked so as to use only the + sign, with inversion signals where necessary, i.e. to change a . sign to a + producing the same function, and where necessary to give a positive output. The first example worked above shows how this is done with an AND gate. NAND and NOR functions can be obtained in a similar way.

The NAND function is already in OR logic:

$$\overline{A} + \overline{B} = S$$

Employment of an OR gate to yield a NAND function theory simply requires inversion of both inputs — Fig. 4.3(a).

The NOR function is in AND logic:

$$\overline{\overline{A}}.\overline{\overline{B}} = S$$
$$\overline{A} + \overline{B} = \overline{S}$$

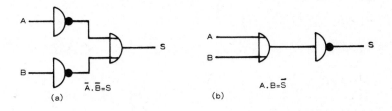

A.$\overline{B}$=S (a)   A.B=$\overline{S}$ (b)

Fig. 4.3(a). OR function performed by two NOT and one OR device.
Fig. 4.3(b). NOR function is performed by OR and NOT devices.

Inversion on inversion puts the equation back to its original state, so this expression simplifies to:

$$A + B = \bar{S}$$

Thus the NOR function is performed in OR logic by an OR gate followed by a NOT gate for inversion — Fig. 4.3(b).

AND *logic*

Here the aim is to express all equations with the . (AND) sign. Obviously an AND gate already does this:

$$A.B = S \text{ (Fig. 4a)}$$

Other logic functions can be determined from an AND gate as follows:

The OR function can be provided by inversion $\bar{A} + \bar{B} = \bar{S}$ (Fig. 4b)

check by inverting again $\qquad \overline{\bar{A} + \bar{B}} = \bar{\bar{S}}$

which is the same as $\qquad A + B = S$

(the OR function)

The NOR function is already in AND logic $(\bar{A}.\bar{B} = S)$ — (Fig. 4c).

The NAND function is devised simply by inversion of the output of an AND gate $A.B = \bar{S}$ (Fig. 4d).

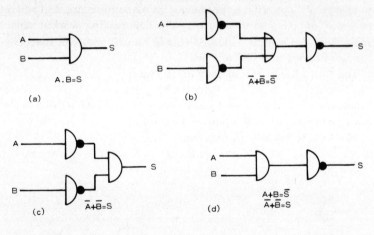

Fig. 4.4. AND, OR, NOR and NAND functions devised from AND logic.

NAND *logic*

Here the requirement is to express all equations in the form $\bar{\bullet}$ (inverted AND).

To derive the OR function $\qquad\qquad A + B = S$

invert $\qquad\qquad\qquad\qquad \bar{A}.\bar{B} = \bar{S}$

invert again	$\overline{\overline{A} + \overline{B}} = \overline{\overline{S}}$
which is the same as	$A + B = S$ (Fig. 4.5a).
To derive the NOR function	$\overline{A}.\overline{B} = S$
invert	$\overline{\overline{A} + \overline{B}} = \overline{S}$
invert again for a positive output	$\overline{\overline{A}.\overline{B}} = \overline{\overline{S}}$
which is the same as	$\overline{A}.\overline{B} = S$ (Fig. 4.5b).
To derive the AND function	$A.B = S$
invert	$\overline{A + B} = \overline{S}$
invert again for a positive output	$\overline{\overline{A}.\overline{B}} = \overline{\overline{S}}$ (Fig. 4.5c).
which is the same as	$A.B = S$

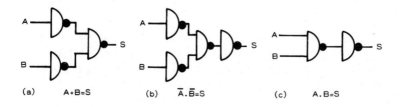

(a)   A+B=S   (b) $\overline{A}.\overline{B}$=S   (c)   A.B=S

Fig. 4.5. OR, NOR and AND functions devised from NAND logic.

EXCLUSIVE OR

The OR gate, described previously, provides an output of one or more inputs has a value of 1. More specifically it can be described as *inclusive* OR. There is a possible variation with a 2-input OR gate where there is an output if one and *only one* of the inputs has a value of 1. This is known as the *exclusive* OR (it is also written XOR, or sometimes described as *non-equivalence*) — shown in Fig. 4.6 — and has the following truth table:

Input		Output
A	B	S
0	0	0
0	1	1
1	0	1
1	1	0

In other words, there is an output of A = 1 or B = 1, but *not* if the values A = 1, B = 1 occur simultaneously.

The corresponding Boolean equation is:

$$(A + B) (\overline{A}.\overline{B}) = S$$
$$\text{or } A\overline{B} + \overline{A}B = S$$

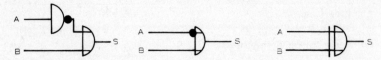

Fig. 4.6. Exclusive OR shown in three different forms.

A particular application of an exclusive OR is as a comparator or equality detector. For example, if the two input signals applied to the gate differ, there will be an output, i.e. the gate 'compares' the two signals and detects a difference. Conversely, if the two input signals are identical the exclusive feature means that there will be no output. This absence of output indicates an *equality* of inputs.

ENABLE

Enable is an inhibit, such as provided by a NOT applied to one input of an AND gate, e.g. as shown in Fig. 4.7 for a 2-input AND gate with inhibit. The third input is called the *strobe* (S) or *enable* input giving the following truth table, where the output is designated Q.

Input			Output (Q)
A	B	S	0
0	0	0	0
0	1	0	0
1	0	0	0
1	1	0	1
0	0	1	0
0	1	1	0
1	0	1	0
1	1	1	0

It will be seen that there is an output (1) *only* when A = 1 and B = 1, *and* S = 0. The presence of an *inhibit* signal (S = 1) holds the output at 0 irrespective of any possible combinations of A and B, i.e. even when A = 1, B = 1.

The corresponding Boolean equation is:

$$A.B.\bar{S} = Q$$

Fig. 4.7. ENABLE has an inhibit function on the S input.

Solving Problems

The basic process of designing logic circuits to meet particular requirements is to break down the problem into elementary 'yes-no' or 'stop-go' steps involving formal logic and co-relating these steps as necessary. In the technical language, this means dealing with original truths (the facts of the question) called *propositions* and putting these together to arrive at an answer or *syllogism* based on the presence of these truths. Specifically, for example, a single truth can be dealt with by NOT logic, i.e. the output responding to an input is either NOT (i.e. not true) or NOT NOT (i.e. true). Normally, however, more than one input is involved and there is some inter-relationship between inputs, calling for the use of *connections* expressing the relationships. The chief of these are AND and OR.

The following is a problem involving several propositions and connections, representing the qualifications necessary to qualify for an executive position, say:

   A. University education.

or B. Technical college training with relevant certificates.

   C. At least five years practical experience in a certain profession.

   D. Over twenty five years of age.

   E. Not married.

In plain language the basic relationship is:

A OR B AND C AND D AND NOT E

The corresponding equation in Boolean algebra is:

$$(A + B).C.D.\bar{E} = S$$

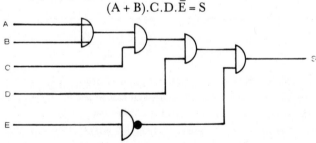

Fig. 4.8. Problem solution using AND, OR and NOT devices.

An immediate solution employing AND, OR and NOT logic elements is shown in block form in Fig. 4.8. This also follows directly from the Boolean equation. Suppose, however, that only AND and NOT devices are available, i.e. the problem must be solved in AND logic. This means that the equation must be rendered in terms of ● only.

Start by inverting the original equation:

$$(\bar{A}.\bar{B}) + \bar{C} + \bar{D} + \bar{\bar{E}} = \bar{S}$$

Now invert again:

$$\overline{(\bar{A}.\bar{B})}.\bar{\bar{C}}.\bar{\bar{D}}.E = \bar{\bar{S}}$$

Note here that by containing $(\bar{A}.\bar{B})$ as one term in a bracket it does not change its state on inversion.

Now remove double inversions as they merely mean using pairs of NOT devices to get back to the original signal:

$$(\overline{\bar{A}.\bar{B}}).C.D.\bar{E} = S$$

The bracketed term $(\overline{\bar{A}.\bar{B}})$ remains as something of a problem as it still contains double inversion. However, since we are restricted to NOT and AND devices, this is really no problem at all as it can be accommodated by a NOT device in each input to an AND, and a further NOT in the output. The final circuit in AND logic is then as in Fig. 4.9.

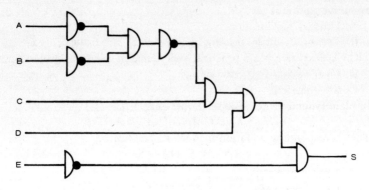

Fig. 4.9. Problem solution using AND and NOT devices.

Given no restrictions on availability of components, then further solutions can be worked in Boolean algebra to see if any simpler circuit can be derived. There is, in fact, using NOR logic ($\bar{+}$).

Starting with	$(A + B).C.D.\bar{E} = S$ again
and inverting	$\overline{(A + B)} + \bar{C} + \bar{D} + \bar{\bar{E}} = \bar{S}$
inverting again as a whole	$\overline{\overline{(A + B)} + \bar{C} + \bar{D} + \bar{\bar{E}}} = \bar{\bar{S}}$
and removing double inversions	$\overline{(A + B)} + \bar{C} + \bar{D} + E = S$

Remembering that bracketed inputs, i.e $(A + B)$ in this example, must be directed to one separate (NOR) device the final circuit then works out as in Fig. 4.10. This saves two components compared with the AND logic circuit of Fig. 4.9.

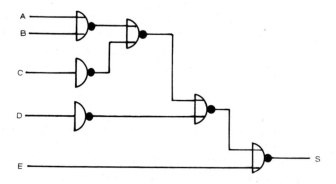

Fig. 4.10. Simpler solution to problem using OR and NOT devices.

Whether solution by Boolean algebra is quicker or simpler than design by block logic diagrams is debatable. For some people it is, for others it is not. Where it does have a definite advantage is in positive elimination of unnecessary components by making it simple to spot and remove double inversions.

Boolean Algebraic Theorems

Most problems can be solved by applying the appropriate Boolean algebra *theorems,* or basic rules under which Boolean algebra works. Only one has been mentioned so far, de Morgan's theorem, which is basically:

$$\overline{A.B.C.} = \bar{A} + \bar{B} + \bar{C}$$
$$\text{or} \quad \overline{A + B + C} = \bar{A}.\bar{B}.\bar{C}$$

There are numerous others — some obvious, others rather more difficult to understand at first. Those which may be of particular significance or use are:

$\bar{\bar{A}} = A$ or $\bar{\bar{B}} = B$, etc. Double inversion returns the function to its original form.

$A.A = A.$ In other words, with an AND device application of the same signal to both inputs will result in the same output.

$A + A = A.$ The same as above, but in this case relating to an OR device.

$A.\bar{A} = O$ or $A.\bar{B} = O.$ With one input inverted there is no output from an AND.

$A + \bar{A} = 1$ or $A + \bar{B} = 1.$ With one input inverted, provided one has a value of 1 there is always an output from an OR.

A number 1 appearing in a Boolean equation means that one signal is always applied; or an 0 that there is always no signal at that particular input. (The numbers 1 or 0 in this case would replace A or B, etc., on a particular input diagram).

It follows that:

A.0 = 0   (the AND function can never be completed with one input always at 0).

A.1 = A   (the AND function is completed with a single input A when the second input is 1, and output is governed by the value of A).

A + 0 = A   (the OR function is complete with one input signal if the other input signal is 0).

A + 1 = 1   (the OR function is complete with a single input when the second input is 1, and output = 1. Compare with the AND equivalent).

Functions enclosed by a bracket are subject to *normal* algebraic treatment when expanded, e.g.

A.(B + C) or A *and* (B *or* C)

    becomes A.B. + A.C (A *and* B *or* A *and* C)

(A + B).(C + D) or (A *or* B) *and* (C *or* D)

    becomes A.C + A.D + B.C + B.D

    (A *and* C *or* A *and* D *or* B *and* C *or* B *and* D)

Checking by writing out in words and comparing with the original expression will verify if the original expansion is correct or not.

A + (A.B) = A. This is self explanatory on spelling it out — A OR A AND B. It is an OR function satisfied if only A is present.

A + ($\bar{A}$.B) = A + B. Again this is an OR function, so it is satisfied if A *or* B are present, viz:

    A OR NOT A AND B = A OR B.

CHAPTER 5

Logic Circuit Devices

Basically all the functions of a logic switching system can be provided by NAND/NOR gates, or alternatively by either an AND or OR gate(s) and inverter(s). The former is the preferred method since AND/OR circuitry has a number of practical limitations. If AND/OR elements are cascaded, for example, each produces some attenuation of the signal which may require the addition of power amplifiers at certain stages, thereby complicating the circuit design. With NAND/NOR circuit design this is not necessary, the main requirement here being merely to observe the maximum number of inputs (fan-in) and outputs (fan-out) provided by each element.

Initially all electronic logic circuits were constructed from discrete components, i.e. transistors and diodes for active elements and resistors and capacitors for passive elements. Typically these yielded printed circuit modules about 1in. to 2in. × 1in. (25-50 × 25mm) for assembly into complete circuits. These have now been (almost) entirely replaced by integrated circuits offering the performance capabilities of numerous interconnected modules in a single miniaturized 'package'. Besides offering very great reductions in weight and size, integrated logic circuits also have the advantages of greater reliability, greater speed of operation and reduced power consumption. Also they are now generally cheaper than all the components needed to construct discrete modules covering the same functions.

Some integrated circuits have the disadvantage of lower signal levels — of the order of 0.8-2 volts as compared with 6-12 volts (or even 24 volts) normally employed with discrete modules. This renders the integrated circuit more susceptible to noise and can place a premium on component location, lead length and earthing requirements. However, the widely used CMOS IC's can be used over a wider voltage range and have very high noise immunity.

As with discrete component modules, integrated logic circuits are based on the same components, e.g. transistors, diodes, etc., although in very much miniaturized form.

Diagramatically, therefore, the two forms of circuits are identical, although for the purpose of use only the external connection points of the integrated circuit normally need to be identified.

The diode-resistor network shown in Fig. 5.1(a) provides *positive* AND logic. With all inputs A,B,C positive (i.e. logic 1) all the diodes are reverse-biased and do not conduct, giving an output of + E (i.e. logic 1). In the absence of any *one* input that diode will conduct, when the output will fall to 0.

The same circuit worked with *negative* logic (i.e. − E corresponding to logic 1) will work as an OR gate giving a 1 output in the presence of any input.

Equally, if the bias voltage is made more positive than logic 1, all diodes will conduct when all the inputs are present together, clamping the output to logic level 1.

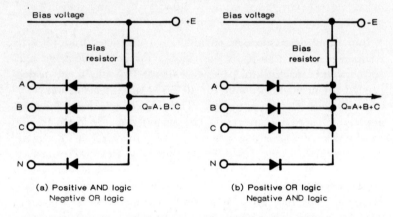

(a) Positive AND logic
Negative OR logic

(b) Positive OR logic
Negative AND logic

Fig. 5.1. Diode-resistor logic networks.

The network shown in Fig. 5.1(b) has the diodes connected the other way round. This time, with *positive* logic (+ E as input) it will work as an OR gate; and with *negative* logic (− E as input) as an AND gate. Again there is the possibility of clamping the output if required.

The disadvantage of these networks is that if the circuits are cascaded the input current to any one circuit must be provided by the circuit preceding it. This means that relatively low values of bias resistors must be used in order to maintain the required drive currents. In practice this may not be possible and buffer amplifiers have to be inserted between stages.

Diode-transistor Logic (DTL)

Diode-transistor logic overcomes this limitation by incorporating a transistor amplifier in the output circuit. A typical positive logic NOR gate of this type is shown — Fig. 5.2. Here any input going positive (logic 1) will cause the base of the transistor to go positive with respect to the emitter and cut off — output will then be logic 0 (no current flow through the collector circuit). When all inputs are logic 0 the base of the transistor will be negative, yielding a collector output approaching the emitter value or logic 1.

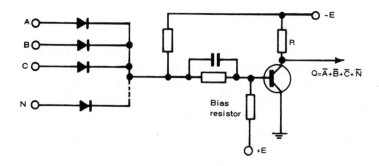

Fig. 5.2. Diode-transistor logic (DTL).

Worked with *negative* logic (– E = logic 1), this circuit provides a NAND function.

DTL logic was originally widely produced in integrated circuit form operating at speeds of 2-20 MHz with logic levels between 0.5 and 5 volts, and for power supplies between 3 and 6 volts. It has now been replaced by simpler and more efficient networks, e.g. transistor-transistor logic (TTL) and more exotic devices.

Resistor-Transistor Logic (RTL)

Resistor-transistor logic is another network form which was widely used for discrete modules, but found less suitable in integrated circuit form because of its low logic levels (about 1 volt) on 3-4 volt supplies. It has also poor fan-out (limited number of outputs) and noise immunity. It is show here in Fig. 5.3 as still being of interest for discrete module construction since it is a simple and straightforward circuit with a wide tolerance for variations in component working values.

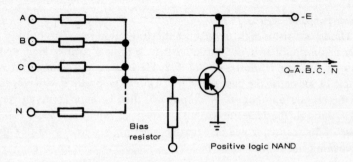

Fig. 5.3. Resistor-transistor logic (RTL).

Direct-Coupled-Transistor Logic (DCTL)

With DCTL logic only transistors are used as the switching elements with the advantage of requiring only one low voltage supply with low power consumption and fast switching speeds. It is attractive for producing integrated circuit NAND and NOR gates utilizing a minimum of components — *see* Fig. 5.4.

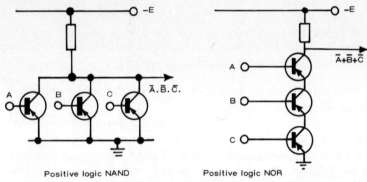

Fig. 5.4. Direct-coupled-transistor logic (DCTL).

Disadvantages of this network are that each input requires its own transistor and these transistors must have uniform characteristics, making it an unattractive choice for construction in discrete modules. These limitations are not so significant in integrated circuit construction, but the type is still relatively susceptible to noise.

Emitter-coupled-Transistor Logic (ECTL)

In this form of direct-coupled transistor network the transistors are not allowed to saturate fully and switch a constant current from one transistor to another. For this reason it is sometimes called Current

Mode Logic (CML). It is considerably less susceptible to noise than DCTL and has much higher switching speeds.

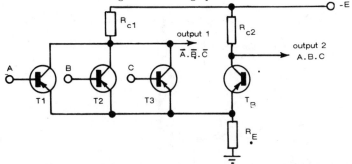

Fig. 5.5. Emitter-coupled transistor logic (ECTL) NAND gate.

Fig. 5.5 shows the network for a NAND gate. Here the bias voltage maintains a constant current through T_B if all the inputs are at a positive level (0V = logic 1). Output at S1 is then negative = logic 0 ($\bar{A}.\bar{B}.\bar{C}$) and positive at S2 = logic 1 (A.B.C). If any input goes negative (– E or logic 0) its transistor will conduct through R_E causing T_B to cut off. In this case output 1 goes to earth (logic 1) and output 2 goes to – E (logic 0).

A feature of this circuit is that it provides a NAND function at output 1 and an AND function at output 2.

Transistor-Transistor Logic (TTL)

In this network transistors are connected in the common base mode, a typical circuit being shown in Fig. 5.6. All NOR inputs have to be negative (logic 0) for the output to go positive (logic 1). Any input giving positive will cause its transistor to conduct, and transistor T_B to cut off. Hence output will then be 0.

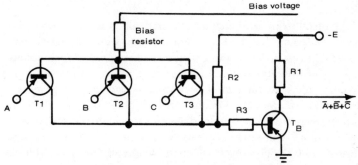

Fig. 5.6. Transistor-transistor logic (TTL).

Rendered in integrated circuit form a multiple-emitter transistor is normally used, with the corresponding circuit shown in Fig. 5.7.

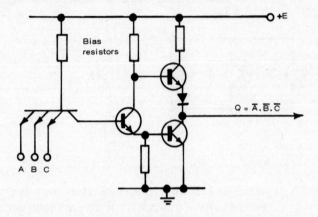

Fig. 5.7. TTL NAND logic.

Circuits of this type are fast switching (e.g. of the order of 4-50 MHz) with good noise immunity and are relatively simple to produce. They are one of the main types used in digital integrated circuits. A typical TTL device can drive up to ten TTL inputs (i.e. has a fan-out of ten), but should not be connected with outputs of different families in parallel unless having a modified output stage.

Most of the range of IC devices in TTL are also produced in low power Schottky logic based on Schottky diodes and Schottky transistors. These have the advantage of faster switching speeds and lower current consumption (only about 25 per cent of the operating level of typical TTL devices).

MOSFET

The metal oxide semi-conductor is basically a special form of field effect transistor (FET), hence the abbreviation MOSFET, often shortened to MOS. It has the attraction of being particularly suitable for extreme miniaturization, i.e. large-scale integration. MOS devices are thus widely used in digital electronics as logic gates, registers and memory arrays. MOSFET circuits consist entirely of FETs (except for parasitic capacitors in certain dynamic applications), but may be made with a Zener diode between the gate and substrate of each (or selected) FET(s). The object of this is to protect the gate from excessive voltages. Under normal operation the Zener diode remains open with no effect in the circuit, but the maximum gate voltage that can arise is limited to the maximum value of the Zener voltage.

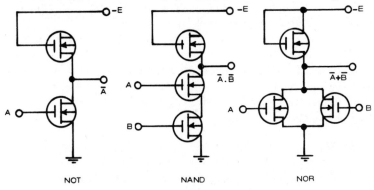

Fig. 5.8. Typical MOSFET gate circuits.

Examples of MOSFET gate circuits are shown in Fig. 5.8, together with a standard circuit symbol for a MOSFET. (There are variations on the symbols used for MOSFETs, e.g. *see* Fig. 5.9).

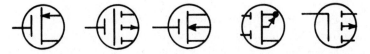

Fig. 5.9. Alternative symbols for MOSFETS. An inward-pointing arrow designates a p-channel MOSFET, with arrow direction reversed for an n-channel MOSFET. The second symbol from the right specifically applies to an enhancement mode device.

MOSFET gates are, in fact, examples of direct-coupled-transistor logic (DCTL). The only basic difference is that because of the high density of components on the same chip it becomes necessary to minimize power consumption in large scale integration although their efficiency, in terms of power performance, is superior to that of ordinary bipolar DCTL gates.

Complementary MOS

Complementary MOS or CMOS employs *complementary* MOS devices on the same chip (i.e. p-channel and n-channel devices). By this means it is possible to reduce power dissipation to very low levels, e.g. as small as 50 nanowatts. Like MOSFET, the basic device is an *inverter*. Combinations of inverters can be used to provide CMOS, NAND and NOR gates, etc.

About the only real disadvantages shown by MOSFET and CMOS devices are their slower speed of working compared with some other devices, and certain high frequency limitations inherent with field effect transistors due to internal capacitance effects.

There are subtle differences between the characteristics of MOSFETS and FETs. The drain resistance of a MOSFET is lower than that of an FET, whilst the resistance between gate and drain or gate and source is higher. In all cases, however, these resistances are extremely high and virtually equivalent to open circuits when shunted by external circuit resistors.

LOCMOS

Local Oxidation Complementary MOS is basically similar to CMOS but with an output buffer stage to improve noise immunity, and provide near ideal transfer characteristics because of the increased voltage gained from 'buffering'. These devices are particularly suitable for pulse shaping since output transistor working is virtually independent of input rise and fall times.

MOS Logic

MOS logic elements are now widely used and have largely taken over from TTL for integrated circuits. The extremely high component density possible means that large memories, shift registers, etc., can be produced in very compact packages. Whilst functions performed are basically similar to those of other logic devices, the behaviour and specific characteristics of MOS and related devices do differ appreciably and need to be appreciated.

The mode of working of *asynchronous* MOS circuits is similar to that of other transistor gates using field-effect transistors (FET's), which differ from bipolar function transistors in being *unipolar,* have a high input resistance, and are generally less 'noisy' than bipolar transistors. Their main disadvantage is the lower gain and the susceptibility of the thin silicon layer of the gate to damage by excessive voltage. MOS field-effect-transistors are also slower working than bipolar transistors.

The majority of such circuits use p-channel enhancement mode MOS devices, where the drain supply is negative potential, and thus work with negative logic. In other words a high negative voltage represents logic '1'.

Supply voltage for such devices commonly range from $-10v$ to $-20v$ with logic '1' having a value of the order of -10 volts. With higher voltages logic '0' then normally lies at a level of – 2.5 to – 5 volts.

P-channel and n-channel MOSFETS may also be used in complementary configuration to operate with *positive* logic. The particular advantage of this is that n-channel devices are faster working and so such circuits can have faster switching times than p-channel devices. Two basic complementary MOS gates are shown in Fig. 5.10.

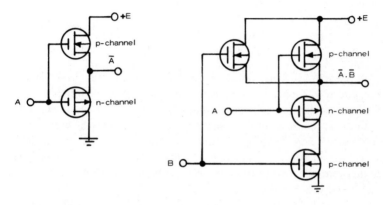

Fig. 5.10. Two basic complementary MOS (CMOS) gates.

Clocked MOS Circuits

MOS circuits are particularly suitable for *synchronous* systems, i.e. clocked circuits, these being generally known as *dynamic* MOS circuits. The advantage here is that average power consumed by the system is reduced. However, where gates are cascaded it is necessary to have more than one pulsed supply to allow for the time it takes the output voltage to reach a steady state. These pulsed voltages are then applied sequentially to the system, giving *two-phase* systems, *three-phase* systems, *four-phase* systems, etc.

Dynamic MOS Inverter

A basic circuit for a *dynamic* MOS *inverter* is shown in Fig. 5.11, operating as mentioned before with negative logic. This requires a train of pulses to operate. At state logic '0' (no pulse) both transistors are switched off and there is minimal power consumed. With the appearance of a negative pulse, both transistors are switched on and conduct with the output being the inversion of the input, i.e. A = 1, Q = 0; or A = 0, Q = 1. The output is held on for the duration of the pulse by the change on the output capacitor C.

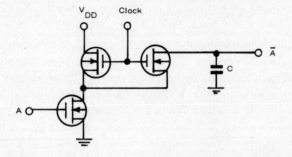

Fig. 5.11 Dynamic MOS inverter.

A particularly important feature of a dynamic MOS circuit is that the parasitic capacitance between gate and substrate inherent in a MOSFET is used to provide temporary memory or storage capacity with a time constant of the order of milliseconds. This storage can be 'refreshed' and made permanent by the application of a clock waveform of suitable frequency, i.e. giving pulse times substantially higher than the time constant of delay. A typical 'refresher' frequency employed is normally 1 kHz or higher.

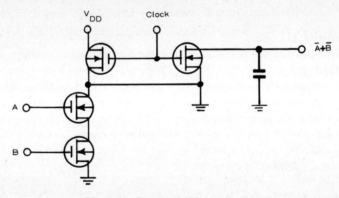

Fig. 5.12. Dynamic MOS NAND gate.

Dynamic MOS NAND gate

A basic circuit for a dynamic MOS NAND gate is shown in Fig. 5.12. This is similar to the *static* NAND gate (Fig. 5.8) except for the additional FET which works as a switching element controlled (i.e. switched on and off, respectively) by the clock pulse. Again in the off condition all transistors are off and power dissipation is minimal.

The *dynamic MOS NOR* gate is again similar to a static NOR gate with an additional FET acting as a switch for the clock pulse — Fig. 5.13.

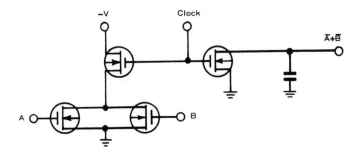

Fig. 5.13. Dynamic MOS NOR gate.

Handling MOS Devices

MOS integrated circuits are more readily damaged than other devices and thus need handling and mounting with care. They are readily damaged by static changes or transient high voltages.

Ideally they should only be handled on a conductive surface, e.g. a metal table top, to which the person handling the device is also connected, e.g. by a metal bracelet, or conductive cord or chain.

Similar recommendations apply when mounting MOS devices on a printed circuit board. If it is impractical to ground the printed circuit board, then the person mounting the circuits should touch the board first to discharge any static before the MOS device is brought into contact with the board.

In practice the most modern CMOS IC's are difficult to damage and the only precaution necessary is to store the chips in conductive plastic carriers, so that all pins are shorted together.

Integrated Circuits and Minimization

The ready availability of complex circuitry in integrated circuit chips has considerably changed attitudes towards circuit design and construction. Medium scale integration (MSI) can offer dozens of gates in a single package; large scale integration (LSI) hundreds of gates in a single chip. The question of *minimization* or the elimination of redundant gates then becomes relatively unimportant. A standard IC 'package' for a computer, decoder, shift register, read-only-memory, etc., may provide more internal circuits than are actually required, but will still offer the most straightforward — and cheapest — solution even if all the pins are not used. ˙

This, too, has influenced design technique itself. Instead of designing a specific, individual circuit as in the days of module construction with discrete components, the circuit designer is more and more having to accept what is predesigned in an IC package and use the facilities it provides accordingly. In other words the designer has to work with *sub-systems,* rather than specific gates or other binary units. This, in fact, has resulted in new design techniques being developed for implementing circuit performance requirements with IC subsystems.

Standard IC Gates

Integrated circuits are produced in a variety of 'packages'. The most common are the TO5 style 'can', similar in size and appearance to a transistor, but with as many as 12 leads emerging from the bottom; and the *flat* package. The latter is of rectangular 'wafer' form; or dual-in-line (DIL) with connections brought out at right angles from both sides — Fig. 5.14. The DIL pack is larger, much easier to mount on printed circuit boards, and also cheaper to produce.

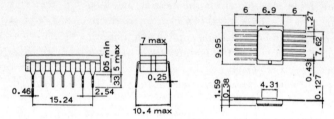

Fig. 5.14. Dual-in-line and flat IC 'packages'.

Common forms of digital IC gates are quadruple two-input NAND, triple three-input NAND, dual four-input NAND, single eight-input NAND, quadruple two-input NOR, quadruple two-input AND, inverters, buffers, etc.; but there are many more.

Such gate circuits are available in most logic families, particularly DTL, TTL, DCTL and ECTL. The limitation on the number of gates per chip is normally set by the number of pins available, e.g. common numbers for a flat package are 14 – , 16 – , and 24 – leads.

Where two different families of ICs may be involved in a complete circuit (e.g. TTL and MOSFET) the question of compatability can arise because of the difference in operating voltage levels. Such differences can be accommodated by buffer circuits 'dropping' a higher level voltage to a lower level voltage where required.

Multiple Gate IC's

Integrated circuits commonly contain multiple circuits or complete sub-systems in a single package, e.g. dual, triple and quadruple gates; hex buffers and inverters; flip-flops and latches; shift registers; counters; multiplexers; mnemonics; display drivers; and arithmetical circuits. All such packages may appear similar, except for the number of leads. The designation of the leads is therefore of primary importance.

Pin numbering reads around the IC, e.g. left to right, then right to left (Fig. 5.15). Note also that some ICs do not have a notch marking pin 1 position but a dot mark instead.

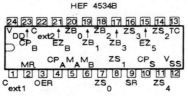

Fig. 5.15. Conventional method of lead or 'pin-out' numbering. This IC is a real time S-decoder counter.

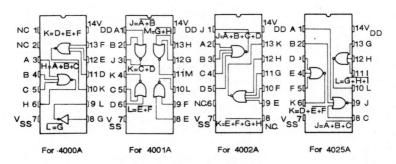

Fig. 5.16. Family of IC NOR gates.

For example, Fig. 5.16 shows a 'family' of NOR gates with the internal devices shown in symbolic form together with their connection to external leads. Externally there is no difference in the appearance of these packages, although they have quite different functions and external connections.

4000A is a dual 3-input NOR gate (two gates) plus inverter

4001A is a quad 2-input NOR gate (four gates)

4002A is a dual 4-input NOR gate (two gates)

4025A is a triple 3-input NOR gate (three gates).

The coding applicable to the connection diagrams is:

A,B,C,D, etc — inputs to gate(s)

J,K,L, etc — outputs from gates

V_{DD}-V_{SS} — supply voltage (Here V_{SS} is the most negative power supply to the device, typically the 'earth').

N.C. — not connected (unused terminals).

This provides all the information necessary to connect the chosen IC into a given circuit. A point to watch here, however, is that although the circuit diagram shows the terminal numbers to which various connections must be made, these will not necessarily follow the same pattern or order of numbers on the actual IC package. The circuit diagram shows connection points — not an actual picture of the IC and its true terminal arrangement.

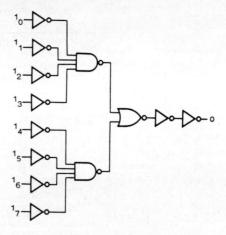

Fig. 5.17. Logic diagram for an 8-input IC NOR gate.

If the circuit is to be *designed* around the IC, then the *electrical characteristics* as specified by the manufacturer need to be known as well. A logic diagram of the IC can also be helpful. Fig. 5.17, for example, is a logic diagram for an 8-input NOR gate IC (HEF 4078B).

IC Buffers

Individual buffer circuits are produced in IC form, the usual number being six contained in a standard 16-pin package. These may be inverting buffers or non-inverting buffers, described as hex inverting buffers (Fig. 5.18) or hex non-inverting buffers (Fig. 5.19), respectively.

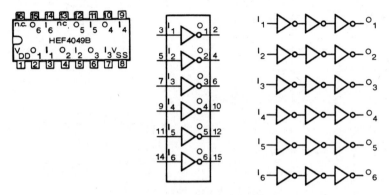

Fig. 5.18. Hex inverting buffer IC.

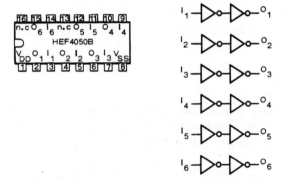

Fig. 5.19. Hex non-inverting buffer IC.

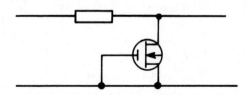

Fig. 5.20. Input protection for buffer circuits.

Where buffers are provided with *input protection* (Fig. 5.20) input voltages in excess of the noted supply voltage for the buffers can be accepted. Such buffers can also be used to convert CMOS logic levels of up to 15 volts to standard TTL levels.

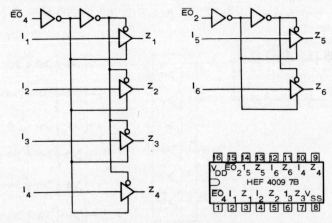

Fig. 5.21. 3-state hex non-inverting buffer.

Hex buffers are also produced with 3-state outputs — Fig. 5.21. Here the 3-state outputs are controlled by two *enable* inputs. A predetermined number of buffers can then be made to assume an off state via the appropriate *enable* signal regardless of the input conditions.

Schmitt Trigger

The Schmitt trigger is another hex (six gate) IC form, this time in 14-pin packages. These trigger circuits are available in inverting and non-inverting forms — Fig. 5.22.

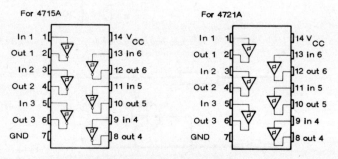

Fig. 5.22. Hex Schmitt trigger IC.

Complex IC's

Integrated circuits embodying complete sub-systems may have 14, 16, 24 or even 28 leads, each are specifically designated. This may be in words and/or code letters. Abbreviations commonly used are:

A0, A1, A2, etc., for inputs (especially address inputs)

01, 02,03, etc., or Q0, Q1, Q2, etc., for outputs

D for data input

E for enable

EL for latch enable

C, Ck, Cp for clock (input)

CE for clock enable

R for reset

S0, S1, S2, etc., for select inputs

ST for strobe input

Cl or CL for clear

R/W for read/write input

Digital Families Compared

DTL, originally widely used for the production of NAND gates, is now largely regarded as obsolescent for IC production. Its chief limitations are that it has limited fan-out and a relatively high propagation delay (typically 30 ns per gate). Only a low voltage supply is necessary, however, and power dissipation is low.

TTL has similar (or slightly higher) power dissipation, but smaller propagation delay and very good noise immunity. MOS and CMOS devices are slower than TTL and also more sensitive to capacitance loading. LOCMOS comes between the two in this latter respect, the buffered outputs reducing the effect of capacitive loading.

CMOS is particularly suited to LSI because of the very small device size possible and the higher potential packing density. LOCMOS and TTL elements are generally produced in SSI and MSI complexity. LOCMOS speed is about three to six times slower than TTL or low-power Schottky (LS-TTL).

In terms of power dissipation LS-TTL and TTL are similar, with MOS lower and CMOS substantially lower. With LOCMOS additional power is required to charge and discharge on-chip capacitances as well as load capacitances, and power dissipation is related to frequency. Starting at very low power (similar to CMOS), the power consumption of a LOCMOS gate generally exceeds that of a LS-TTL between 500 kHz and 2 MHz of actual output frequency; and standard TTL between 1 MHz and 10 MHz. Such a comparison is not necessarily realistic, however, for in a complex circuit only a small fraction of the gates actually switch at the full clock frequency. Most gates operate at a much lower frequency and thus consume less power in the case of a LOCMOS device.

Some comparative data are summarized in the following table:

	DTL	Standard TTL	MOS	Low power Schottky	CMOS	LOCMOS
Supply voltage	low	low	15 - 20	-	1.5 - 18	3 - 15
Power dissipation per gate mW	8 - 12	12 - 22	0.2 - 10		0.1 - 1	varies with frequency
Quiescent power mW		10				
Propagation delay ns	30	10	300	10	70	15 - 40*
Clock frequency MHz	8	35	2	45	5	20 - 8*
Noise immunity	fair to good	very good	low	very good	very good	good
Fan out	8	10	20	10	50	50

* dependent on voltage

CHAPTER 6

Flip-Flops and Memories

The flip-flop is a basic digital 1-bit memory circuit, now commonly called a *memory* device (in logic terms). Originally (and then invariably) explained in the form of a two-transistor circuit with cross-linked resistors and capacitors, it is more simply illustrated and described in terms of logic elements, i.e. two NOT gates (single-input NAND gates) cross-coupled as shown in Fig. 6.1.

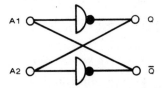

Fig. 6.1. Flip-flop circuit from two cross-coupled NOT gates.

In this configuration the output of each gate is connected to the input of the other. It can exist only in one of two stable states, either $Q = 0$, or $Q = 1$ (with $\bar{Q} = 1$ or $\bar{Q} = 0$, respectively). It can thus 'hold' or latch on to 1-bit of memory, and for this reason is also referred to as a *latch*.

To work as a memory device it is necessary to add a means of *setting* an input, and also a reset or *clear* the memory. This requires the use of two-input NAND gates with one input of each cross-limited to the output of the other gate; and the other input of each gate accepting 'command' signals S (for set) and R (for clear or reset) passed through NOT gates — Fig. 6.2.

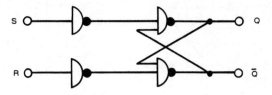

Fig. 6.2. Memory device with set and clear (S – R flip-flop).

In this configuration $S = 1$, $R = 0$ provides $Q = 1$ stored in the

memory. Similarly S = 0, R = 1 provides Q = 0, either resetting from a '1' state or to store 0 in the memory.

This describes the *S – R flip-flop* which has two stable operating states, triggered from one state to the other by a control signal. It can be further modified by 'gating' or *clocking* the input signals, as shown in Fig. 6.3. The main significance of this circuit is that the memory will not accept any information *unless* the clock signal is present.

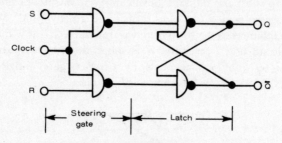

Fig. 6.3. S-R flip-flop with clocked input.

The truth table for an S – R flip-flop is:

S	R	Q_{n+1}
0	0	Q_n
1	0	1
0	1	0
1	1	ambiguous

This illustrates a basic limitation of the S – R flip-flop. Should signal conditions S = 1, R = 1 occur, the memory could be 1 or 0. This can be overcome by further modification to the basic flip-flop circuitry. There are three further variations used — the J – K flip-flop, the T flip-flop and the D flip-flop.

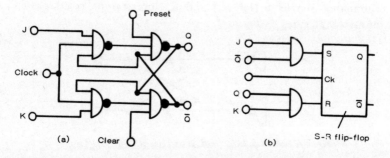

Fig. 6.4. J – K flip-flop configurations.

J – K Flip-Flop

The *J – K flip-flop* aims at removing the ambiguity in the truth table either by the addition of an extra input to each input side NAND gate to accept feedback (Fig. 6.4); or by the addition of a two-input AND gate to the S and R inputs (Fig. 6.4b). Either will render the S = 1, R = 1 condition *determinate*, i.e. Qn + 1. However, working in this mode is still time-dependent and ambiguity can still occur, especially in IC circuits with small propagation times.

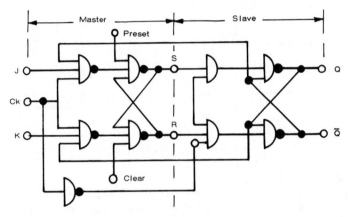

Fig. 6.5. J – K master-slave flip-flop.

Master-Slave J – K Flip-Flop

The *J – K master-slave* flip-flop uses two S – R flip-flops in cascade with feedback from the second or *slave* to the input of the first or *master* — Fig. 6.5. Positive clock pulses are used to excite the master and these are inverted before exciting the slave. This inhibits any possibility of an output oscillating between 0 and 1 in a J = 1, K = 1 condition. The only requirement now is that both the J and K values must remain constant for the pulse duration.

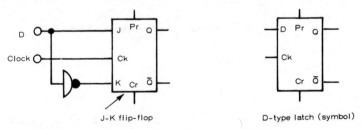

Fig. 6.6. D-type flip-flop or latch.

D-type Flip-Flop

The D-type flip-flop is simply a J – K flip-flop with the addition of an inverter in the K line so that K is the complement of J — Fig. 6.6. There is no ambiguous state since the condition J = K = 1 is not possible. Also the bit on the D line is transferred to the output on the *next* clock pulse, introducing *delay*. It is because of this characteristic that it is often used as a buffer store between a counter and a digital read-out device to introduce delay or eliminate flicker whilst the counter is operating, as well as a basic *latch*.

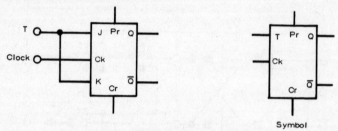

Fig. 6.7. T-type flip-flop or toggle.

T-type Flip-Flop

The *T-type* flip-flop works as a 'trigger' or toggle switch, changing its state with each clock pulse. It has a single input line T (Fig. 6.7). Output remains unchanged at either 0 or 1 levels, so that $Qn + 1 = Qn$; or when the input changes from 0 to 1. Input changes from 1 to 0, however, invert the output so that $Qn + 1 = \bar{Q}n$. When this occurs, the flip-flop is said to *toggle*.

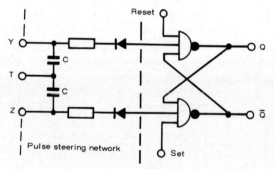

Fig. 6.8. Unclocked T-type flip-flop with pulse steering network.

The T-type flip-flop was also originally produced in unclocked form, comprising two gates giving set-reset stages together with a pulse steering network — Fig. 6.8. It still has attractions for discrete module construction, but is severely limited in respect of count rates possible. For this reason the unclocked T-type flip-flop is not usually produced in IC form.

Summary

The $J - K$ flip-flop is probably the most versatile of all the types, with its truth table generally similar to that of an $S - R$ flip-flop. The behaviour of D-*type* and T-*type* flip-flops is more limited as the following table shows:

Flip–Flop Truth Tables

S – R			J – K			D–type		T–type	
Sn	Rn	Qn+1	Jn	Kn	Qn+1	Dn	Qn+1	Tn	Qn+1
0	0	Qn	0	0	Qn	1	1	1	$\bar{Q}n$
1	0	1	1	0	1	0	0	0	Qn
0	1	0	0	1	0				
1	1	?	1	1	$\bar{Q}n$				

Sample-and-Hold

A sample-and-hold circuit is an *analogue memory*, a basic circuit for which is shown in Fig. 6.9. A negative sampling pulse applied to the gate closes the circuit allowing the capacitor to change to the instantaneous voltage of the input. In the absence of a pulse the gate circuit opens with the capacitor retaining its charge. The output is thus a steady voltage level charging in steps between the sampling pulse intervals.

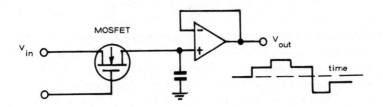

Fig. 6.9. Sample-and-hold or analogue memory.

To work effectively the time of the sampling pulses must be short, the value of the capacitor low and the output impedance of the op-amp high in order not to discharge the capacitor between sampling pulses. Also the capacitor must be of a type which can hold its full charge between sampling pulses.

A field-effect transistor (MOSFET) is preferable to a bipolar transistor switch in most sample-and-hold applications, although the latter can be used.

Read-only Memory (ROM)

Read-only memory or ROM is a circuit which accepts a binary code (known as an *address*) at its input terminals and provides another binary code or *word* at its output terminals for each of the input combinations. Basically, therefore, it is a code-conversion system, although essentially it consists of a *decoder* applied to the input signals feeding an *encoder* providing the output signals. Since this encoder is essentially a *memory matrix* the information it is provided with is stored and can be read out as often as required — hence the description read-only memory (Fig. 6.10).

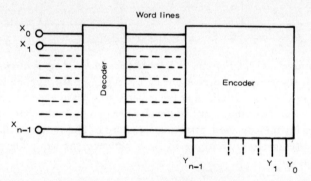

Fig. 6.10. Read-only memory (ROM).

Specifically a ROM will have a specified number of inputs X_0, X_1, X_2, etc., and a specified number of outputs Y_0, Y_1, Y_2, etc. These numbers will not necessarily be the same. Thus if there are X inputs and Y outputs, the capacity of that ROM is 2^X words each of Y bits, or a $2^X \times Y$-bit memory. For example, if there are 32 inputs with 8 outputs this particular ROM would have a capacity of 32 words each of 8 bits, or a $32 \times 8 = 256$-bit memory. ROMs from 256-bit up to 1024-bit are typical figures for MSI using COSMOS in conjunction with TTL logic. With LSI much larger bit memories can be achieved in a single package. Alternatively ROMs can be cascaded to provide larger memories.

The way a ROM works is to decode the input into *word lines* W_0, W_1, etc., which are the *minterm* outputs of the decoder. These lines are then encoded again in the memory matrix where they are held. The working relationship can be established by a truth table or Boolean equations, or both, as a guide to implementation.

Taking a 4-input 4-output ROM (64-bit memory) as a simple example, the truth table for conversion from binary code to a Gray code would be:

binary inputs				Word line	Gray code outputs			
X3	X2	X1	X0		Y3	Y2	Y1	Y0
0	0	0	0	W0	0	0	0	0
0	0	0	1	W1	0	0	0	1
0	0	1	0	W2	0	0	1	1
0	0	1	1	W3	0	0	1	0
0	1	0	0	W4	0	1	1	0
0	1	0	1	W5	0	1	1	1
0	1	1	0	W6	0	1	0	1
0	1	1	1	W7	0	1	0	0
1	0	0	0	W8	1	1	0	0
1	0	0	1	W9	1	1	0	1
1	0	1	0	W10	1	1	1	1
1	0	1	1	W11	1	1	1	0
1	1	0	0	W12	1	0	1	0
1	1	0	1	W13	1	0	1	1
1	1	1	0	W14	1	0	0	1
1	1	1	1	W15	1	0	0	0

and the corresponding Boolean equations would be:-

☐ Y0 = W1 + W2 + W5 + W6 + W9 + W10 + W13 + W14

☐ Y1 = W2 + W3 + W4 + W5 + W10 + W11 + W12 + W13

☐ Y2 = W4 + W5 + W6 + W7 + W8 + W9 + W10 + W11

☐ Y3 = W8 + W9 + W10 + W11 + W12 + W13 + W14 + W15

To accommodate different arithmetic codes some IC ROMs are designed to be *programmable* after manufacture. Such a device is called a PROM.

Random-Access Memory (RAM)

A *random-access memory* or RAM is a similar device to a ROM except that the stored words may be addressed and written directly as well as being read. The decoder in this case employs *latches* (flip-flops) instead of diodes or transistors, which are bistable devices. This means that whilst a RAM provides stored memory, this is lost when the power supply is removed. For this reason a RAM is described as a *volatile* device, with power dissipation necessary to maintain storage. In the case of certain types, e.g. a dynamic MOS RAM, a 'refreshing' charge is necessary at regular intervals (e.g. every millisecond or so) to replace leakage of all capacitance on which the memory depends.

Dynamic MOS RAM

In a dynamic MOS RAM information can be stored on the parasitic gate-to-substrate capacitance, resulting in considerable circuit simplification, e.g. only three devices are needed to store 4 bits instead of the eight required in a static MOS RAM. In this case, however, 'refreshing' of all bits is required.

Typical IC RAM

The Siemens 5607 MOC IC is a 10K-bit static RAM with inputs compatible with TTL. It is shown in physical (28-pin flat package) and block diagram forms in Fig. 6.11. Access line is approximately 10ηs and maximum power dissipation 700 mW.

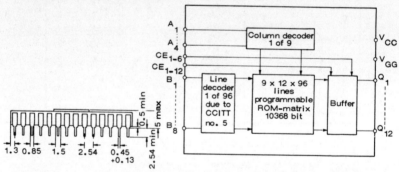

Fig. 6.11. Static 10 K bit IC ROM (S607).

IC RAM

Fig. 6.12 shows the physical form (14-pin flat package) and block diagram for a 64-bit, 1-bit per word random access read/write memory.

The memory is strobed for reading or writing only when the strobe input (ST), chip enable inputs (CE_1 and CE_2) are HIGH simultaneously. The output data is available at the data output (D_{OUT}) only when the memory is strobed, the read/write input ($R/\overline{W}$) is HIGH and after the read access time has passed. Note that the output is initially disabled and always goes to the LOW state before data is valid. The output is disabled when the memory is not strobed or $R/\overline{W}$

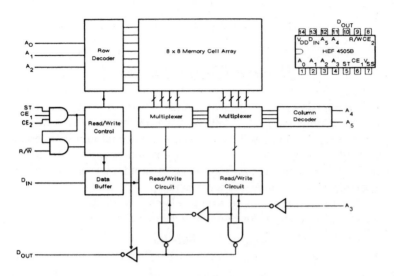

Fig. 6.12. 64-bit IC RAM (HEF4505B)

is LOW. R/W̄ may remain HIGH during a read cycle or LOW during a write cycle. The output data has the same polarity as the input data. *The function table is:*

ST, CE_1, CE_2	R/W̄	D out	mode
LOW	LOW	floating	disabled
HIGH	LOW	floating	enabled (write)
LOW	HIGH	floating	disabled
H	H	memory data	enabled (read)

Registers

Flip-flops are a binary device and thus have a memory capacity of 1 bit (i.e. are a 1-bit memory). It follows that a combination of flip-flops can store as many bits as there are flip-flops, i.e. N flip-flops can store an N-bit word. Such a combination of flip-flops or binary memory devices is called a *register*. Normally to allow the word data to be fed in serially, flip-flops are connected serially, i.e. output to input. The data is then progressively shifted along the line of flip-flops to complete the word. In this case the circuit is referred to specifically as a *shift register*.

These are described in more detail in Chapter 12.

CHAPTER 7

Binary Coded Decimals

Whilst digital electronic elements 'think', 'count' or react in terms of binary arithmetic (0 or 1, 'off' or 'on'), the human brain finds it much easier to think and communicate in decimal numbers. Some method of being able to render binary numbers in easily readable decimal equivalents is therefore highly desirable, i.e. an in-between system presenting binary coded decimals.

This can be done quite simply. To represent the ten decimal numbers from 0 to 9, four binary digits or *bits* are required, viz:

decimal	pure binary			
	(2^3)	(2^2)	(2^1)	(2^0)
0	0	0	0	0
1	0	0	0	1
2	0	0	1	0
3	0	0	1	1
4	0	1	0	0
5	0	1	0	1
6	0	1	1	0
7	0	1	1	1
8	1	0	0	0
9	1	0	0	1

It is readily possible to write decimal number equivalents of 0 to 9 in separate *groups* of four bits, using as many groups as necessary to cover the number of digits in the decimal number. Taking decimal number 7,893 as an example and treating each digit separately as a number between 0 and 9:

decimal	7	8	9	3
binary coded decimal	0111	1000	1001	0011

This works equally easily the other way round. To translate a binary coded decimal into its decimal equivalent each group is connected in turn, e.g.

binary coded decimal	0101	0011	1000	0111
decimal	5	3	8	7

i.e. 5,387

This particular system is known as an 8421 binary coded decimal, or 8421 BCD. The numbers here actually refer to the assigned values or *weights* given to the respective groups. Using four groups, as in the example, the weights are:

$$2^3 = 8 \qquad 2^2 = 4 \qquad 2^1 = 2 \qquad 2^0 = 1$$

A little further study will show that with this method of grouping the four *bits* actually provides 16 possible combinations, only ten of which are used to cover the decimal numbers 0 to 9. In other words six of the combinations are redundant, or unnecessary.

In a practical application, say using memory gates or *flip-flops*, to locate decimal numbers when fed by the 8421 BCD code (or vice versa), each group of numbers would need four flip-flops, and each set of four groups would have six unnecessary (redundant) code combinations. These could be eliminated by the use of a (suitable) alternative BCD.

In fact there are many possible BCD code sequences, the relative advantages of each depending on a variety of factors such as simplicity of circuit construction, operating speed and ease of decoding for read-out purposes. Some are *weighted* codes, whilst others are not.

Basic requirements of a weighted code are that the weights must be chosen so that their number is not greater than 15 and not less than 9. Additionally, one of the weights must be 1, and another either 1 or 2. For example some possible combinations are (omitting those of little or no use):

8421 BCD (already described) 7421, 5421, 5211, 2421.

Here are the respective group equivalents:

decimal	pure binary				binary coded decimal			
	(2^3)	(2^2)	(2^1)	(2^0)	7421	5421	5211	2421
0	0	0	0	0	0000	0000	0000	0000
1	0	0	0	1	0001	0001	0001	0001
2	0	0	1	0	0010	0010	0100	0010
3	0	0	1	1	0011	0011	0110	0011
4	0	1	0	0	0100	0100	0111	0100
5	0	1	0	1	0101	1000	1000	0101
6	0	1	1	0	0110	1001	1001	0110
7	0	1	1	1	1000	1010	1011	0111
8	1	0	0	0	1001	1011	1110	1110
9	1	0	0	1	1010	1100	1111	1111

The 7421 BCD code has a particular advantage in practical application in that it employs a minimum number of 1's. The figure 1 in a binary device represents an 'on' state, normally drawing current. Thus this code is attractive for providing minimum current consumption.

5211 BCD and 2421 BCD (or any other code where the sum of the weights is 9), yield the property that the 9's complement of the number (i.e. 9 – N. where N is the number) can be obtained simply by inverting the binary equivalent. For example, in 5211 BCD, decimal 6 is given by 0.1001. Inverting this gives 0110 or decimal 3 (i.e. 9 – 6 = 3). This again can be of particular advantage for certain workings.

Three other codes are worth mentioning here (although there are many more). These are the 'excess three' code, the reflected or Gray code, and the Johnson code.

The *excess three* is a self-complementing code obtained by adding 3 to each group of the binary code. It is very useful for performing decimal or binary coded decimal arithmetic.

The reflected binary or *Gray* code is also widely used, particularly in digital shaft position encoders as it incurs only one digit change in passing from any one combination to the next.

The *Johnson* code is quite different as this is an unweighted code, particularly adapted to counting because of the simplicity with which it can be decoded into decimal.

The table below shows the equivalents in the three codes for decimals 0-9:

decimal	pure binary				excess three code	Gray code	Johnson code
	(2^3)	(2^2)	(2^1)	(2^0)			
0	0	0	0	0	0011	0000	00000
1	0	0	0	1	0100	0001	00001
2	0	0	1	0	0101	0011	00011
3	0	0	1	1	0110	0010	00111
4	0	1	0	0	0111	0110	01111
5	0	1	0	1	1000	0111	11111
6	0	1	1	0	1001	0101	11110
7	0	1	1	1	1010	0100	11100
8	1	0	0	0	1011	1100	11000
9	1	0	0	1	1100	1101	10000

Error-detecting codes

Where the code used contains *redundancies* the appearance of a redundancy number would indicate an error. For example, the appearance of 1111 when using 8421 BCD would indicate an error since no such number exists in the code. Errors produced by dropping or gaining a digit in the same code group, however, would not be apparent as they still show valid combinations. The same is true of all codes used having no redundancies.

The simplest method of error-detection is to add an extra bit, called a *parity bit* in each group, giving this a value of 0 or 1 to make the total number of 1's in each group either odd or even. Should an error occur this will immediately show up by the fact that the number of digits in the group will no longer be odd (or even).

The limitation of this is that only single errors will show up, e.g. *two* errors occurring in the same group would return the sum of digits to odd (or even) and show as correct. *Three* errors in the same group would again indicate an error — but not whether a single or triple error.

To check blocks of information the read-out can be arranged in the form of a matrix. Parity checks are then made on the rows and columns, including the extra row formed by the column parity check (which also needs its own parity bit).
Example (odd parity):

(decimal 8732)   10000   01110   00111   00100
Note. The last digit is the *parity bit* in these examples:

decimal	BCD	parity bit	total bits
8	= ←1000	0	odd, OK
7	= ←0011	1	odd, OK
3	= ←0011	1	odd, OK
2	= ←0010	0	odd, OK

Rather more elaborate codes may be used in a similar way, e.g. 'N-out-of-M' codes where the presence of an error gives rise to a non-valid combination; or the *Diamond* code. The former will only detect single errors. The Diamond code is designed to detect multiple errors, using the property of all numbers which obey the formula $3n + 2$. The check is made by subtracting 2 from the combination and dividing the remainder by binary 3. If there is no remainder, the combination is valid.

Examples of error-detecting codes

decimal	PB	2^3	2^2	2^1	2^0	2-out-of-5 code	Diamond code
		odd parity bit					
0	1	0	0	0	0	01100	00010
1	0	0	0	0	1	10001	00101
2	0	0	0	1	0	10010	01000
3	1	0	0	1	1	00011	01011
4	0	0	1	0	0	10100	01110
5	1	0	1	0	1	00101	10001
6	1	0	1	1	0	00110	10100
7	0	0	1	1	1	11000	10111
8	0	1	0	0	0	01001	11010
9	1	1	0	0	1	01010	11101

Octal Numbers

The octal system is a numbering system to base 8. In other words it has eight digits, from 0 to 7, relative to the decimal system, although decimal 10 equals octal 8.

The advantage of octal numbers is that they can be written as groups of three binary digits, called binary triplets. Thus, conversion from a binary number to an octal number is direct and straightforward:

Octal	Binary Triplet		
0	0	0	0
1	0	0	1
2	0	1	0
3	0	1	1
4	1	0	0
5	1	0	1
6	1	1	0
7	1	1	1

To convert a binary number into its octal number the binary number is broken down into groups of three or triplets. If necessary, zeros are added in front of the number to complete a set of triplets. The corresponding octal number then follows from the equivalent of the various triplets.

Example: Binary 10110011

group in triplets 10 110 011

add zero to complete 010 110 011

corresponding octal numbers 2 6 3

i.e. octal number = 263.

Octal numbers can be used to check computer arithmetical solutions by comparing the answers obtained by the two numbering systems. A worked-out example should make this clear.

Binary sum	Octal sum
110	6
+ 010	+ 2
1000	10 octal

001 000

1   0   octal eqivalent

The two octal numbers agree, i.e. the one derived directly by octal number working and the other extracted as the octal equivalent of the binary sum solution. Thus, the binary arithmetic is correct.

Handling Fractions

In the decimal system fractions are, of course, simply designated by a decimal point. Fractions are thus expressed in negative base values, i.e. 10^{-1}, 10^{-2}, 10^{-3}, etc.

Exactly the same principle applies with any other numbering system although the resulting fractions will have quite different values. In the case of the *binary* system, for example, the negative base values are 2^{-1}, 2^{-2}, 2^{-3}, etc., the corresponding fractions being ½, ¼, ⅛, etc. Here are some typical comparisons:

	decimal				*binary*			
	10^0	10^{-1}	10^{-2}	10^{-3}	2^0	2^{-1}	2^{-2}	2^{-3}
(⅛)	0 .	1	2	5	0 .	0	0	1
(¼)	0 .	2	5	0	0 .	0	1	0
(½)	0 .	5	0	0	0 .	1	0	0
(¾)	0 .	7	5	0	0 .	1	1	0
(1)	1 .	0	0	0	1 .	0	0	0

	octal			*duo-decimal*		
	8^0	8^{-1}	8^{-2}	12^0	12^{-1}	12^{-2}
		⅛	1/64		1/12	1/144
(⅛)	0 .	1	0	0 .	1	6
(¼)	0 .	2	0	0 .	3	0
(½)	0 .	4	0	0 .	6	0
(¾)	0 .	6	0	0 .	9	0
(1)	1 .	0	0	1 .	0	0

One point which arises is that to express a fraction *exactly* the denominator of the fraction must be exactly divisible by the base of the system. Thus the binary system can accommodate all fractions whose denominator is divisible by 2. It cannot, for example, accommodate $\frac{1}{3}$ as an *exact* value; nor can the decimal system ($\frac{1}{3} = 0.3$ recurring). The duo-decimal system, being of base 12, which is divisible by 3, can. Here $\frac{1}{3} = 0.4$.

CHAPTER 8

Clocks

Devices where the output is a function of both present and past inputs occurring in sequence require some form of 'timing'. They are then generally described as *synchronous* circuits (or *clocked* or *strobed* circuits), controlled by a fundamental clock frequency of the system. In digital circuits the clock is generally some form of *pulses* or square waves, generated by a bistable unit or incorporating a clock input into a basic switching circuit.

Basically, therefore, a clock as used in digital circuits is an oscillator which generates square waves or pulses (unlike a 'radio' oscillator which generates sine waves). This is a function readily performed by an op-amp combined with an integrator as shown in Fig. 8.1.

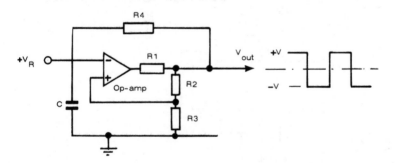

Fig. 8.1. Simple square wave generator.

In this circuit the output is either $+ V$ or $- V$. The op-amp works as a *comparator*, comparing the input voltage V_1 with a standard reference voltage V_R, V_1 being in the form of feedback from the voltage divider provided by R2, R3. If V_1 is positive, then $V_o = - V$. The capacitor C then changes to $+ V$, when after a period of time the comparator output reverses and the capacitor charges to $- V$. The result is a square wave output with a time interval determined by the values of R4 and C (the integrator part of the circuit). In practice maximum pulse frequency obtainable from such a basic circuit is of the order of 10 kHz.

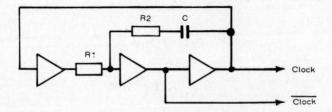

Fig. 8.2. Simple practical circuit for square wave generator using L29016 op-amps.
If R2 = 100 ohms for frequency 10 kHz,   C = 0.18 μF
                                 100 kHz, C = 0.018 μF
                                 1 MHz, C = 1600 pF
                                 5 MHz, C = 200 pF
Value of R1 is 300 ohms.

A practical circuit of this type is shown in Fig. 8.2, based on a standard IC hex inverter (but using only three of the inverters). Frequency is governed by the values of R2 and C used and can range from 10 kHz to 5 MHz.

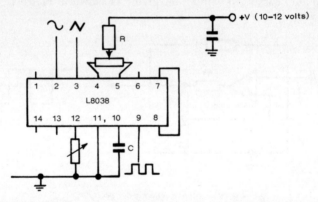

Fig. 8.3. IC waveform generator providing sine, triangular or square wave outputs.

Some IC waveform generators provide square, triangular and sine wave outputs simultaneously, one example being the L8038. It can also be phase locked to a reference. A working circuit for this IC is shown in Fig. 8.3. Square wave amplitude is of the order of 0.9V.

The frequency is set by R and C, viz:

$$\text{frequency} = \frac{0.15}{R \times C}$$

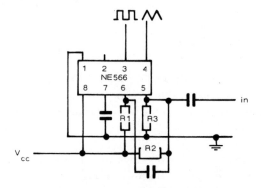

Fig. 8.4. IC waveform generator providing square or triangular waveform outputs.

An alternative IC which provides both square and triangular waveform outputs is shown in Fig. 8.4. Here the frequency is given by:

$$\text{frequency} = \frac{2(+V - V_{CC})}{R1 \times C1 \times V_{CC}}$$

Simple Multivibrators

Multivibrators are *analogue* rather than digital devices, but they are readily capable of working as pulse generators. A *monostable* multivibrator has one stable state and one quasi-stable state. Starting in its stable state, a triggering signal will transform it into its quasi-stable state when after a certain period of time the circuit returns to its stable state. Thus the output is in the form of a pulse with a pulse width equal to the circuit delay time. A further triggering signal is the η necessary to generate another pulse, and so on. Because of this mode of working it is known as a *one-shot multivibrator*.

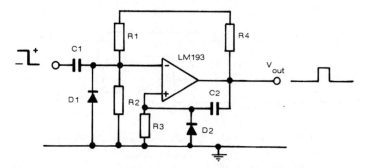

Fig. 8.5. Practical one-shot multivibrator circuit based on LM193 op-amp.
Component values for 1 kHz frequency

R1, R2, R3 — 1 M ohm
R4 — 10 k
C1 — 100 pF
C2 — 0.001μF

A practical circuit is shown in Fig. 8.5. The component values specified giving a pulse width of approximately 1 millisecond. The triggering signal is applied to the non-inverting terminal. The diode D2 acts as a clamp.

A further circuit using two op-amps and capable of generating both positive and negative pulses is shown in Fig. 8.6.

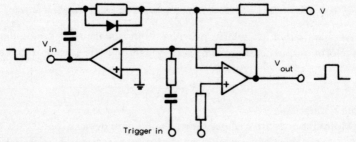

Fig. 8.6. Multivibrator circuit capable of generating both positive and negative pulses.

Bistable Multivibrators

A *bistable multivibrator* is stable in both of its states and is generally known as a *flip-flop*. It is a true digital rather than analogue device, which in a sequential circuit is set and reset by clock pulses. An example of a practical bistable multivibrator circuit is shown in Fig. 8.7 based on an op-amp and five resistors.

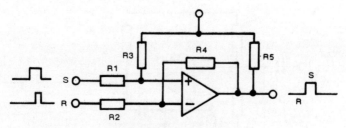

Fig. 8.7. Bistable (digital) multivibrator using LM193 co-amp
Typical component values:
R1, R2, R3 — 100 k
R4 — 50 k
R5 — 10 k

Flip-flops are described in further detail in Chapter 6.

Crystal Controlled Oscillators

A basic circuit for a crystal controlled oscillator is shown in Fig. 8.8. Pulse frequency is now fixed by the frequency of the crystal employed. A rather more practical circuit of this type is shown in Fig. 8.9.

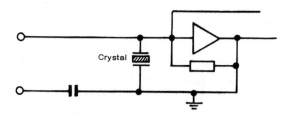

Fig. 8.8. Crystal controlled oscillator circuit.

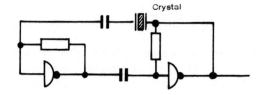

Fig. 8.9. Practical crystal controlled oscillator circuit.

SWEEP GENERATORS

A linear amplifier (op-amp) used in conjunction with a resistor and capacitor can be made to work as an *integrator*. If the input is a constant voltage, the output is in the form of a linear ramp or *sweep* waveform, Fig. 8.10. This is a *linear* device, but may be used in hybrid circuits, so is worthy of brief description.

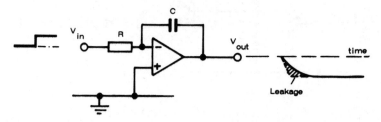

Fig. 8.10. Basic linear ramp or sweep waveform generator.

The *Miller sweep generator* uses bipolar transistors or FETs for the amplifier circuit. Input is in the form of a pulse (trigger) signal, yielding an output of the form shown in Fig. 8.11. The slope towards the bottoming voltage after the initial drop is defined by V/RC, where V is the applied (power) voltage.

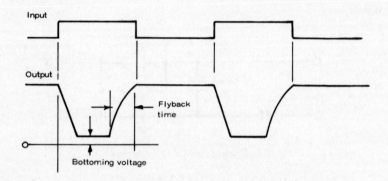

Fig. 8.11. Characteristic output of a Miller sweep generator.

The *bootstrap* sweep generator provides a triangular output waveform, tuned by R1 C1 in the opposite sense, i.e. in the time between trigger pulses. This waveform is the result of bootstrap feedback through C2, Fig. 8.12. A noteworthy point is that if the diode is replaced by a resistor the flyback is prolonged, as shown by the dotted line on the output waveform.

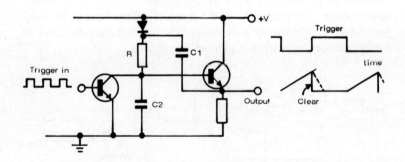

Fig. 8.12. Bootstrap sweep generator.

Schmitt Trigger

The *Schmitt trigger* is another linear device which is essentially a voltage level discriminator. Voltage output is low (but not zero) when the input voltage is low. When the input voltage reaches a pre-determined (design) value the output voltage changes virtually instantaneously to a high value. It is thus a useful triggering circuit and widely used as such. A simple form of this circuit is shown in Fig. 8.13, together with input and output characteristics.

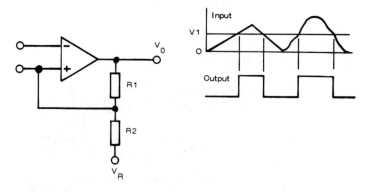

Fig. 8.13. *Basic Schmitt trigger circuit and input and output characteristics. R1, R2 is a potential divider giving a feedback factor of R2/(R2 + R2).*

The Schmitt trigger is widely used to generate trigger pulses of sharp form (i.e. rapid rise and fall times, a basic circuit being shown in Fig. 8.14).

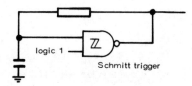

Fig. 8.14. *Schmitt trigger pulse circuit.*

IC *Digital Clock*

The Siemens SAJ 341A 24-pin MOS integrated circuit provides a complete clock function with outputs of minutes and hours. This, in fact, is only part of the complete IC. It also contains a 4-decade up-counter with preset and presettable predividers, as the following shows (*see also* Fig. 8.15):

	$1p_3$	$1p_2$	$1p_1$	function	
1	H	H	L	divide 1:1	⎫
2	H	L	H	10:1	⎪
3	L	H	H	100:1	⎬ counter operation
4	L	H	H	1000:1	⎪
5	L	H	L	6000:1	⎭
6	L	L	H	time base 50 Hz	⎫
7	L	L	L	time base 60 Hz	⎬ clock operation
0	H	H	H	time base 100 Hz	⎭

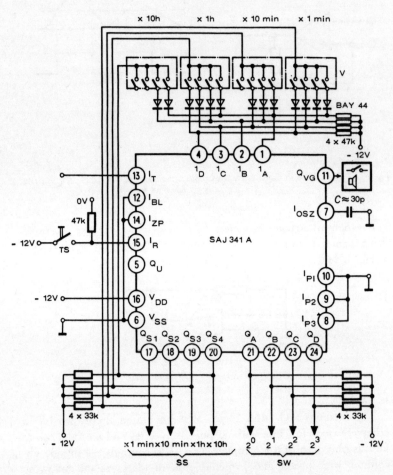

Fig. 8.15. IC digital clock circuit.

Input $1p_1$ HIGH with the other two inputs LOW provides clock operation with a time reference frequency of 50 Hz (standard mains frequency in the UK) applied to the clock input. The counter and divider then change their logic condition with the LOW-HIGH transition of the clock signal.

Clock operation is possible with or without using the pre-select switches. Without preset all preset inputs are set to LOW. By connecting inputs Izp and I_B with bounce-free keys, minute and hour-counters can be set. By applying a LOW signal, single pulses can be applied through Izp to the hour-counter, and through I_B to the minute counter. A defined output position is then achieved by turning on the mains voltage or applying a reset signal to I_R.

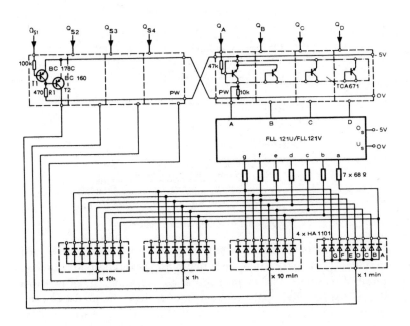

Fig. 8.16. Interface circuit for digital clock.

Using the presets, two further functions can be provided:
(i) Using a key-contact at reset input I_R the clock can be set to a pre-selected time by applying a short signal.
(ii) If provision is made that at a preset time a HIGH signal is present at the comparator output (Q_{VG}) for 1 minute, an additional switching function is available, e.g. to sound an alarm.
The carry output (Q_V) provides a day-carry from 23.59 to 0 hours.
An interface circuit for the SAJ 341A clock with LED display is shown. in Fig. 8.16.

Pushbutton Dialler
Another interesting IC clock circuit is shown in Fig. 8.17. This is described as a *pushbutton dialler*.

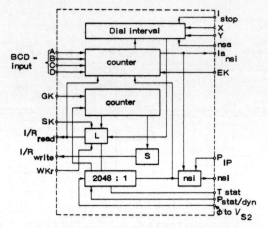

Fig. 8.17. IC Pushbutton-dialler.

At the inputs the circuit is BCD coded. The following circuit then provides clock generation and dial-pulse generation for indirect number selection. Pin 24 provides a stop function. If set to HIGH during the Nth digit, pulse sequences for N digits are produced completely, with further outputs blocked. When 1 stop reacts to LOW, remaining digits are generated after a time delay until the memory is empty. Clock frequency range is 10 kHz to 50 kHz.

CHAPTER 9

Encoders and Decoders

A binary *encoder* consists of a suitable number of inputs, each of which represents a *line* in the binary code involved. It then provides direct access to any one line whereby an input signal applied to that line gives a 1 output, or *generates* a bit.

Suppose the binary code has to cover a 'count' of 10 decimal, i.e. is required to have 10 bits. This can only be satisfied with a minimum of $2^4 = 16$ bits ($2^3 = 8$ is not enough), of which $16 - 10 = 6$ will be redundant, since only 10 lines are required. These are:

decimal		bit 3 (2^3)	bit 2 (2^2)	bit 1 (2^1)	bit 0 (2^0)
		Output Code			
line 0	— 0 —	Y3	Y2	Y1	Y0
line 1	— 1 —	0	0	0	1
line 2	— 2 —	0	0	1	0
line 3	— 3 —	0	0	1	1
line 4	— 4 —	0	1	0	0
line 5	— 5 —	0	1	0	1
line 6	— 6 —	0	1	1	0
line 7	— 7 —	0	1	1	1
line 8	— 8 —	1	0	0	0
line 9	— 9 —	1	0	0	1

This can be encoded in the form of a *diode matrix*, which for simplicity is shown as a wired circuit with keys for each input, with outputs Y0, Y1, Y2, Y3 indicating the *state* of the matrix via lamps — Fig. 9.1. To complete any line circuit to its corresponding lamp current must flow through a diode to provide OR logic. In the absence of a diode in the circuit cleared by any key the corresponding vertical or *output* line will be at signal 0. (This only occurs at key 1 position, representing decimal 0). Equally, each diode can be replaced by a transistor working as a diode (base and emitter connections, with the advantage that only one multiple-emitter transistor is required instead of fifteen diodes. In practice several transistors may be needed for coverage, depending on the number of bits in the output code. The number of emitters required is equal to the number of bits in the code.

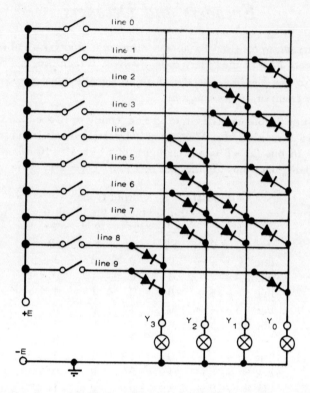

Fig. 9.1. Diode matrix encoder.

Assuming that the keys are not operated simultaneously, operation of a single key will encode the decimal number position in binary equivalent, all other lines at this time being in 'open' state. For example, closing key 7 (to encode decimal 8), output lines Y2, Y1 and Y0 are actuated (through the diodes) giving a complete output signal 0111.

Specifically, Y0 = 1 if line 1, line 3, line 5, line 7 or line 9 = 1. Similarly Y1 = 1 if line 2, line 3, line 6 or line 7 = 1, and so on.

A complete truth table is thus:

inputs (lines)										outputs			
9	8	7	6	5	4	3	2	1	0	Y3	Y2	Y1	Y0
0	0	0	0	0	0	0	0	0	1	0	0	0	0
0	0	0	0	0	0	0	0	1	0	0	0	0	1
0	0	0	0	0	0	0	1	0	0	0	0	1	0
0	0	0	0	0	0	1	0	0	0	0	0	1	1
0	0	0	0	0	1	0	0	0	0	0	1	0	0
0	0	0	0	1	0	0	0	0	0	0	1	0	1
0	0	0	1	0	0	0	0	0	0	0	1	1	0
0	0	1	0	0	0	0	0	0	0	0	1	1	1
0	1	0	0	0	0	0	0	0	0	1	0	0	0
1	0	0	0	0	0	0	0	0	0	1	0	0	1

Expressed in Boolean algebra (remembering that ' + ' means OR logic):

$$Y0 = W1 + W3 + W5 + W7 + W9$$
$$Y1 = W2 + W3 + W6 + W7$$
$$Y2 = W4 + W5 + W6 + W7$$
$$Y3 = W8 + W9$$

Such an encoding matrix, therefore, can be implemented with OR gates and diodes.

DECODERS

A *decoder* is a system whereby digital information is extracted in a different form, e.g. a binary code to be 'read' in decimal equivalent (BCD-to-decimal decoder). Again assuming that the binary unit is a 4-bit device (i.e. with a count of decimal 10) a basic decoder to cover

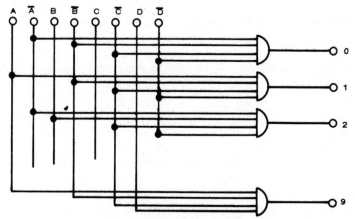

Fig. 9.2. BCD to decimal decoder.

this will require four inputs A, B, C and D — and ten output lines (covering decimal 0 to 9). This can be specifically described as a 4 (bit)-to-10-line decoder.

To accommodate all possible input states eight inputs are required A, $\bar{A}$, B, $\bar{B}$, C, $\bar{C}$, D, $\bar{D}$. To cover ten output lines, ten 4-input NAND gates are needed. The basic circuit is then as shown in Fig. 9.2. (In practice the complementary inputs $\bar{A}$, $\bar{B}$, $\bar{C}$ and $\bar{D}$ may be obtained via inverters). This circuit then works in the opposite-hand sense to a decoder, i.e. outputs and inputs transposed. The truth table is thus:

Inputs					Output Lines									
A	B	C	D		9	8	7	6	5	4	3	2	1	0
0	0	0	0		0	0	0	0	0	0	0	0	0	1
0	0	0	1		0	0	0	0	0	0	0	0	1	0
0	0	1	0		0	0	0	0	0	0	0	1	0	0
0	0	1	1		0	0	0	0	0	0	1	0	0	0
0	1	0	0		0	0	0	0	0	1	0	0	0	0
0	1	0	1		0	0	0	0	1	0	0	0	0	0
0	1	1	0		0	0	0	1	0	0	0	0	0	0
0	1	1	1		0	0	1	0	0	0	0	0	0	0
1	0	0	0		0	1	0	0	0	0	0	0	0	0
1	0	0	1		1	0	0	0	0	0	0	0	0	0

For example, a binary input $\bar{A}B\bar{C}D$ or 0101 gives an immediate output on line 5 (decimal 5).

These requirements can also be implemented by a *diode matrix* working with AND logic.

MULTIPLEXERS

A *multiplexer* provides the facility of selecting 1 out of any number of input sources, directing this data to a single information channel. It is normally specified by an N-to-1 multiplexer, N being the number of inputs it is designed to select from. A typical basic circuit for a 4-to-1 multiplexer is shown in Fig. 9.3 using AND gates and AND-OR logic.

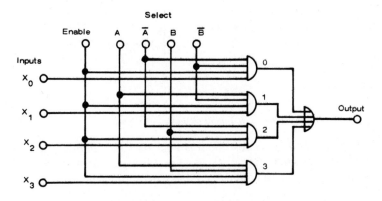

Fig. 9.3. Basic 4-to-1 multiplexer circuit.

DEMULTIPLEXERS

A *demultiplexer* performs the inverse function of a multiplexer, i.e. provides a binary signal on any one of N lines to which it is addressed. It can be derived directly from a decoder by the addition of a signal (S) line — Fig. 9.4. When a date signal is applied at S the output will appear only on the addressed line as the complement of this signal.

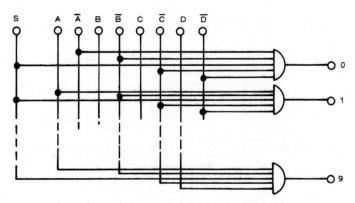

Fig. 9.4. Basic demultiplexer with S input.

In practice this working is normally combined with an inhibit or *enable* input (also called a *strobe* input) feeding the S terminal — Fig. 9.5.

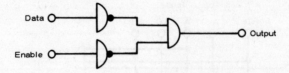

Fig. 9.5. Inhibit or enable input function.

In this case if the enable input is 1 the data are inhibited from appearing on any line. If both data and enable inputs are 0, the data will appear directly on the addressed line without inversion.

The capacity of demultiplexers is specified in the same way as for multiplexers, e.g. 2-to-4 line, 3-to-8 line, 4-to-6 line. The latter is the normal limit for single IC packings.

IC *Decoders*

A practical example of an IC decoder is shown in Fig. 9.6. This has four inputs to accept a 4-bit binary coded decimal (1-2-4-8 BCD code) and 10 outputs 0_0, 0_1, etc. The truth table, written in terms of

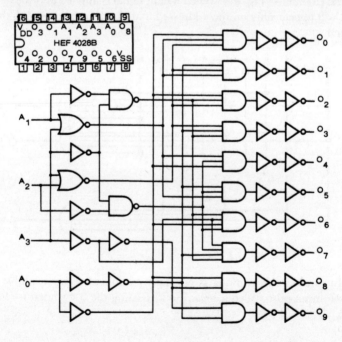

H = high state or more positive voltage signal and L = low state or less positive voltage, is:

inputs				outputs									
A_3	A_2	A_1	A_0	O_0	O_1	O_2	O_3	O_4	O_5	O_6	O_7	O_8	O_9
L	L	L	L	H	L	L	L	L	L	L	L	L	L
L	L	L	H	L	H	L	L	L	L	L	L	L	L
L	L	H	L	L	L	H	L	L	L	L	L	L	L
L	L	H	H	L	L	L	H	L	L	L	L	L	L
L	H	L	L	L	L	L	L	H	L	L	L	L	L
L	H	L	H	L	L	L	L	L	H	L	L	L	L
L	H	H	L	L	L	L	L	L	L	H	L	L	L
L	H	H	H	L	L	L	L	L	L	L	H	L	L
H	L	L	L	L	L	L	L	L	L	L	L	H	L
H	L	L	H	L	L	L	L	L	L	L	L	L	H
H	L	H	L	L	L	L	L	L	L	L	L	H	L
H	L	H	H	L	L	L	L	L	L	L	L	L	H
H	H	L	L	L	L	L	L	L	L	L	L	H	L
H	H	L	H	L	L	L	L	L	L	L	L	L	H
H	H	H	L	L	L	L	L	L	L	L	L	H	L
H	H	H	H	L	L	L	L	L	L	L	L	L	H

(last six rows braced with **X**)

H = HIGH state (the more positive voltage)

L = LOW state (the less positive voltage)

X = state immaterial or 'don't care'

Basically a 1-2-4-8 BCD code applied to the inputs causes the selected output to be H, and the other source L. The device can also be used as a 1-of-8 decoder with enable. In this case 3-bit octal inputs are applied to A_0, A_1 and A_2, selecting an output from O_0 to O_7. Input A_3 then becomes an active LOW enable forcing the selected output to L when A_3 is H.

1-of-16 decoder/demultiplexer

The HEF4515B 1-of-16 decoder/demultiplexer is an excellent example of how much logic can be contained in a small IC package. This has four binary weighted address inputs (A_0, A_1, A_2 and A_3) and 16 outputs, a latch enable input (EL) and an active LOW enable input $\bar{E}$. When EL is HIGH the selected output is determined by the data on A_0 to A_3. When EL goes LOW the last data present are stored (latched) and the outputs remain stable. When $\bar{E}$ goes LOW the selected output is determined by the contents if the latch is LOW and when $\bar{E}$ goes HIGH all outputs are HIGH. The enable input $\bar{E}$ does not affect the state of the latch.

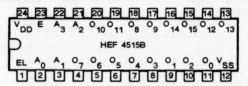

Fig. 9.7. 1-of-16 decoder/demultiplexer.

The logic diagram for this device is shown in Fig. 9.7. The corresponding truth table (EL HIGH) is:

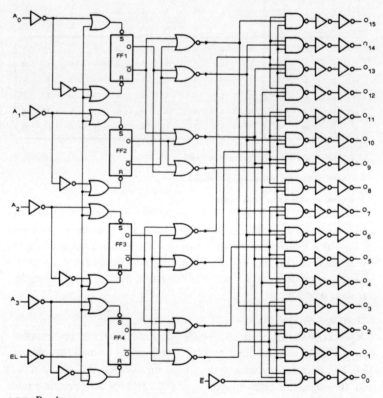

LED *Read-out*

LED displays are commonly used to provide visual read-out of digital information in decimal numbers. Each number requires a seven-segment light-emitting diode (LED) to cover numbers from 0 to

9 — *see* Fig. 9.8. Thus it is necessary to convert digital input to binary coded decimal form into a 7-bit (7-segment) display code, when the following truth table applies:

decimal number	D3	D2	D1	D0	word line	Y6 (g)	Y5 (f)	Y4 (e)	Y3 (d)	Y2 (c)	Y1 (b)	Y0 (a)
								7-bit output code				
0	0	0	0	0	W0	0	1	1	1	1	1	1
1	0	0	0	1	W1	0	0	0	0	1	1	0
2	0	0	1	0	W2	1	0	1	1	0	1	1
3	0	0	1	1	W3	1	0	0	1	1	1	1
4	0	1	0	0	W4	1	1	0	0	1	1	0
5	0	1	0	1	W5	1	1	0	1	1	0	1
6	0	1	1	0	W6	1	1	1	1	1	0	0
7	0	1	1	1	W7	0	0	0	0	1	1	1
8	1	0	0	0	W8	1	1	1	1	1	1	1
9	1	0	0	1	W9	1	1	0	0	1	1	1

Check example: to display decimal 6 the binary code 0110 has to be converted into output code 1111100, powering segments Y6, Y5, Y4, Y3 and Y2, with segments Y1 and Y0 'off'.

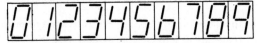

Fig. 9.8. Seven-segment LED display

The simple answer is to use a ROM as a code converter, specifically providing the required output code. LEDs draw only tiny currents and so can be powered directly from the IC output. Since 4 inputs are required the ROM would actually provide 16 possible input combinations, 6 of which are unused. It is possible to design a *minimized* converter circuit with no redundancies and such devices are available in IC form specifically for such applications. They are called N-segment/decoder/drivers, where N is the number of LED segments covered.

LED displays are themselves produced in packaged form, normally with 14 pins to fit standard sockets. Some examples of internal wiring are shown in Fig. 9.9. These are 7-segment *digit* displays plus a decimal point.

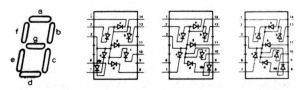

Fig. 9.9. Typical internal wiring of seven-segment LED displays.

Series of such displays, i.e. to read out more than one digit, has a common connection (common cathode or common anode), when the basic circuitry involved is as shown in Fig. 9.10.

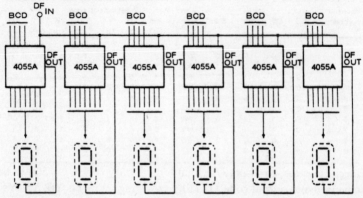

Fig. 9.10. Series connection of LED displays with common cathode (or common anode).

Arrays of this type, of course, are not restricted to numerals. They can present read-out in letters, e.g. A,B,C,D,E,F,G,H, etc., mixed figures and numbers, or other symbols, although available combinations with a 7-segment LED will not cover a full alphabet.

Exactly the same type of device can also be used to power liquid crystal displays.

Display Drivers

An example of an LED display driver is shown in Fig. 9.11 which is in the form of a 16-lead flat package IC. This has four address inputs (coded D_A to D_D) and seven outputs (coded 0_a to 0_g). In addition this is an active LOW latch enable input ($\overline{EL}$) and an active LOW ripple blanking input ($\overline{BI}$), an active LOW lamp test input ($\overline{LT}$).

Fig. 9.11. 16-lead flat package IC LED display driver HEF4511B.

When $\overline{EL}$ is LOW, the state of the segment outputs (0_a to 0_g) is determined by the data on D_A to D_D. When $\overline{EL}$ goes HIGH, the last data present on D_A to D_D are stored in the latches and the segment outputs remain stable. When $\overline{\text{LT}}$ is LOW, all the segment outputs are HIGH independent of all other input conditions. With $\overline{\text{LT}}$ HIGH, a LOW on $\overline{\text{BI}}$ forces all segment outputs LOW. The inputs $\overline{\text{LT}}$ and $\overline{\text{BI}}$ do not affect the latch circuit.

In this description HIGH corresponds to a signal 1 and LOW to a signal 0. The full *function table* (i.e. the truth table plus the other input functions) is then as follows. Input conditions marked X indicate that the state is immaterial or 'don't care'.

	inputs							outputs							
$\overline{\text{EL}}$	$\overline{\text{BI}}$	$\overline{\text{LT}}$	D_D	D_C	D_B	D_A	O_a	O_b	O_c	O_d	O_e	O_f	O_g	display
X	X	L	X	X	X	X	H	H	H	H	H	H	H	8
X	L	H	X	X	X	X	L	L	L	L	L	L	L	blank
L	H	H	L	L	L	L	H	H	H	H	H	H	L	0
L	H	H	L	L	L	H	L	H	H	L	L	L	L	1
L	H	H	L	L	H	L	H	H	L	H	H	L	H	2
L	H	H	L	L	H	H	H	H	H	H	L	L	H	3
L	H	H	L	H	L	L	L	H	H	L	L	H	H	4
L	H	H	L	H	L	H	H	L	H	H	L	H	H	5
L	H	H	L	H	H	L	L	L	H	H	H	H	H	6
L	H	H	L	H	H	H	H	H	H	L	L	L	L	7
L	H	H	H	L	L	L	H	H	H	H	H	H	H	8
L	H	H	H	L	L	H	H	H	H	L	L	H	H	9
L	H	H	H	L	H	L	L	L	L	L	L	L	L	blank
L	H	H	H	L	H	H	L	L	L	L	L	L	L	blank
L	H	H	H	H	L	L	L	L	L	L	L	L	L	blank
L	H	H	H	H	L	H	L	L	L	L	L	L	L	blank
L	H	H	H	H	H	L	L	L	L	L	L	L	L	blank
L	H	H	H	H	H	H	L	L	L	L	L	L	L	blank

H = HIGH state (the more positive voltage)

L = LOW state (the less positive voltage)

X = state is immaterial or 'don't care'

The corresponding *logic* diagram is shown in Fig. 9.12 (p.94).

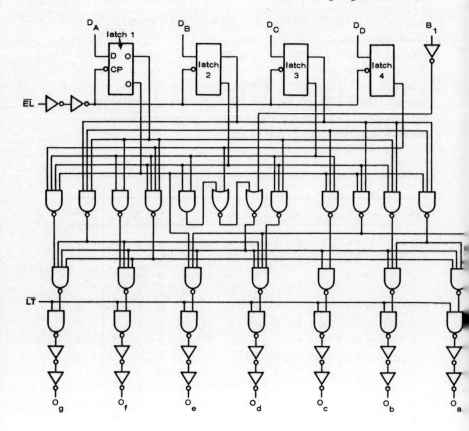

Fig. 9.12. Logic diagram of typical IC display driver (HEF4511B).

CHAPTER 10

Digital Adders

Binary adders perform the mathematical operation of addition using *bits* (binary digits). They can also be used to perform subtraction (negative addition), multiplication (repeated addition) and division (repeated subtraction) by suitable programming. In other words all the common arithmetical functions can be performed by binary adders, which in turn are a basic application of *logic gates*.

Starting with a two-input device, known as a half-adder (HA), this has to cope with $2^2 = 4$ possible combinations of input signals and provide a realistic output. This means coverage of all input conditions in a meaningful way. To do this it must have two outputs — one to provide a read-out for the 'addition' within the capability of a two-digit count (0 and 1), and the other to accommodate 'overflow' or 'carry' to another counting stage.

Calling the inputs A and B and the outputs R (signal read-out or display) and C (carry), the truth table is:

input		sum		output	
A	B	—		R	C
0	0	0	0	0	0
0	1	0	1	1	0
1	0	0	1	1	0
1	1	1	1	0	1

Note that whilst three of the combinations producing the sum can be represented by a single digit read-out either as 0 or 1, the condition 11 cannot. It represents an 'overflow' condition, hence the read-out must revert to 0 with *carry 1*.

In terms of logic gates the first three combinations can be covered by an exclusive OR gate. To accommodate 'carry' an AND gate must be added — *see* Fig. 10.1(a). Fig. 10.1(b) shows the standard symbol for a half-adder.

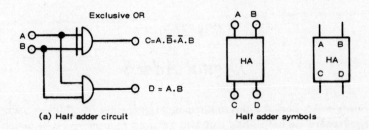

Fig. 10.1. Half adder circuit and symbols.

To extend addition to accommodate more digits (e.g. starting by accommodating the 'carry' from a half adder), half adders can be cascaded, as in Fig. 10.2(a) to make a *full adder*. In this purely diagrammatic form this leaves a spare input to accept another digit directly, i.e. providing a three-input device. A more realistic configuration for a full adder (FA) is as shown in Fig. 10.2(b), with provision to accept two inputs A and B directly into this stage and a 'carry' input from the initial stage (which only need be a half adder). To provide the facility to carry C1 *or* C2 forward to the next stage (there cannot be a carry output at *both* C1 and C2 simultaneously) the 'carry' output must be taken through an OR gate.

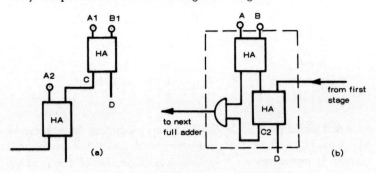

Fig. 10.2. Full adder circuit from half adders.

In practice full adders are not necessarily constructed from two half adders. The number of components required to produce the required function can be reduced since only seven signal combinations are required, as defined by the truth table:

inputs			outputs	
A	B	C in	C out	S
0	0	0	0	0
0	0	1	1	0
0	1	0	1	0
0	1	1	0	1
1	0	0	1	0
1	0	1	0	0
1	1	0	0	1
1	1	1	1	1

The Boolean equations corresponding to the truth table are:

$$Sn = \bar{A}n.Bn.C_{n-1} + \bar{A}n.Bn.\bar{C}_{n-1} + An.\bar{B}n.\bar{C}_{n-1} + An.Bn.C_{n-1}$$
$$Cn = \bar{A}n.Bn.C_{n-1} + An.\bar{B}n.C_{n+1} + An.Bn.\bar{C}_{n-1} + An.Bn.C_{n-1}$$

These equations represent a 'sum of products' and hence each term in the equation is a *minterm*. Considered as minterms, the equations can readily be simplified to:

$$Sn = An.\bar{C}n + Bn.\bar{C}n + An.Bn.C_{n-1} + C_{n-1}.\bar{C}n$$
$$Cn = An.Bn + Bn.C_{n-1} + An.C_{n-1}$$

An example of implementing these simplified equations in hardware form using AND and OR gates is shown in Fig. 10.3.

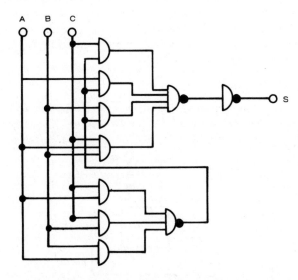

Fig. 10.3. Gate circuit for full adder based on minterms.

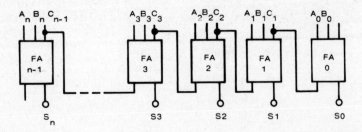

Fig. 10.4. Cascaded full adders.

Fig. 10.4 then shows a 4-bit full adder capable of reading (or displaying) up to a maximum count of $2^3 = 8$ in binary numbers. Note that the working is from right to left, but appropriate inputs can be made to any stage directly, i.e. they do all have to be fed through the first stage as a series of '1' signals. The circuit is an *adder*, not a counter. Also the first (right-hand) stage does not have to be a full adder, only a half adder (although in a practical IC it would usually be a full adder with a third input not used).

Obviously the coverage can be extended by adding further full adders to the left. Commercial IC binary adders are generally available with 1-bit, 2-bit and 4-bit coverage (sometimes more), depending on the number of pins available. A 4-bit adder would require 16 pins — 8 for inputs; 4 for sum outputs; 1 for carry output and 1 for carry input (to allow this IC to be cascade with other full adders); 1 for power input; and 1 for earth connection. Carry connections would be completed internally.

Binary Subtractors

The basic rule of binary *subtraction* is to *add* the (binary) complement of the number to be subtracted. In practice this will involve an extra bit being introduced which may be subject to 'end carry round'.

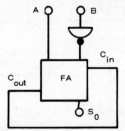

Fig. 10.5. Basic circuit for binary subtraction.

For example, to subtract a 4-bit number from another 4-bit number A the solution is to add A, $\bar{B}$ and 1. The basic circuitry, as applied to a 4-bit adder to turn it into a subtractor is as shown in Fig. 10.5. The basic functions involved for a 4-bit subtractor are:

B plus $\bar{B}$ = 1111

B plus $\bar{B}$ plus 1 = 10000

Hence B = 10000 minus $\bar{B}$ minus 1

When A minus B = A plus$\bar{B}$ minus 1000

The 1 is the output carry C_{out} fed back to the carry input C_{in}.

This works as long as A is greater than B, yielding a positive difference. If B is greater than A, yielding a negative difference, there will be no carry round and a slightly different system must be used. In practice an IC adder/subtractor would incorporate a true/complement unit to handle both positive and negative differences, e.g. *see* Fig.10.6. In the case of a negative difference, the correct solution is then obtained by complementing the sum digits S0, S1, S2 and S3. In the case of a positive difference there is a carry and the solution is given directly by the S0, S1, S2, S3 bits.

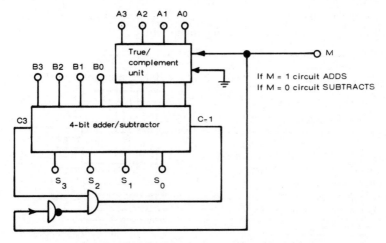

Fig. 10.6. IC adder/subtractor with true/complement unit.

Serial Adder/Subtractors

In the case of a *serial adder* the inputs are synchronous pulse trains applied to individual lines. The output is then either the combined waveform of the inputs (addition) or the difference. This can be performed by a single full adder, with carry facility for subtraction and a time delay in the carry line to inject the carry pulse (when present) into the digit pulses at the correct time interval — *see* Fig. 10.7.

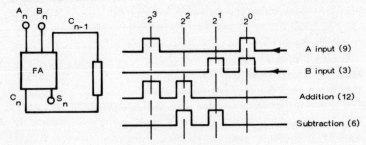

Fig. 10.7. Series adder/subtractor.

The chief advantage of a serial adder/subtractor is that only a minimum of components are required (only one FA and a time delay). It is slower working than the previous type described, which uses *parallel* working, but at the expense of requiring one full adder for each bit.

Various other types of circuits may be employed for adders, e.g. binary-coded decimals, 2's complement, etc.

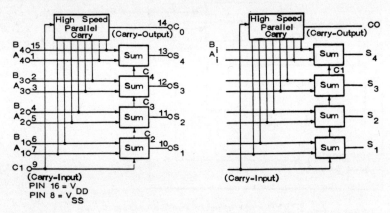

Fig. 10.8. Typical IC adder connection diagram (left) and logic diagram (right).

IC *Adders*

Fig. 10.8 shows connection and logic diagrams for an IC 4-bit full-adder with parallel carry output (HBC/HBS 4008A). It consists of four full-adder stages with the circuitry designed for fast operation.

The basic truth table relating is:

A1	B1	C1	CO	SUM
L	L	L	L	L
H	L	L	L	H
L	H	L	L	H
H	H	L	H	L
L	L	H	L	H
H	L	H	H	L
L	H	H	H	L
H	H	H	H	H

H = high signal
L = low signal

To extend capacity to 8-bits, 12-bits, 16-bits, etc, individual adders can be cascaded.

inputs	outputs															
$\bar{E}\ A_0\ A_1\ A_2\ A_3$	O_0	O_1	O_2	O_3	O_4	O_5	O_6	O_7	O_8	O_9	O_{10}	O_{11}	O_{12}	O_{13}	O_{14}	O_{15}
H X X X X	H	H	H	H	H	H	H	H	H	H	H	H	H	H	H	H
L L L L L	L	H	H	H	H	H	H	H	H	H	H	H	H	H	H	H
L H L L L	H	L	H	H	H	H	H	H	H	H	H	H	H	H	H	H
L L H L L	H	H	L	H	H	H	H	H	H	H	H	H	H	H	H	H
L H H L L	H	H	H	L	H	H	H	H	H	H	H	H	H	H	H	H
L L L H L	H	H	H	H	L	H	H	H	H	H	H	H	H	H	H	H
L H L H L	H	H	H	H	H	L	H	H	H	H	H	H	H	H	H	H
L L H H L	H	H	H	H	H	H	L	H	H	H	H	H	H	H	H	H
L H H H L	H	H	H	H	H	H	H	L	H	H	H	H	H	H	H	H
L L L L H	H	H	H	H	H	H	H	H	L	H	H	H	H	H	H	H
L H L L H	H	H	H	H	H	H	H	H	H	L	H	H	H	H	H	H
L L H L H	H	H	H	H	H	H	H	H	H	H	L	H	H	H	H	H
L H H L H	H	H	H	H	H	H	H	H	H	H	H	L	H	H	H	H
L L L H H	H	H	H	H	H	H	H	H	H	H	H	H	L	H	H	H
L H L H H	H	H	H	H	H	H	H	H	H	H	H	H	H	L	H	H
L L H H H	H	H	H	H	H	H	H	H	H	H	H	H	H	H	L	H
L H H H H	H	H	H	H	H	H	H	H	H	H	H	H	H	H	H	L

H = HIGH state (the more positive voltage)

L = LOW state (the less positive voltage)

X = state is immaterial or 'don't care'

CHAPTER 11

Binary Counters

Counter circuits may be *asynchronous* or *synchronous*, the main difference being that with asynchronous counters all operations (except 'clear') are initiated by the incoming pulses, whereas with synchronous working a separate clock pulse is employed to synchronize operations. Synchronous counter circuits are more complicated to design, generally use more components, but are usually faster in operation.

The basic element employed in a binary counter is a two-state (bistable) electrical device which is either off (0) or on (1), such as a *flip-flop* (FF). A simple element of this type provides a count of 2^0 (= decimal 1). Counting range can be extended by connecting a number of units in series, any 'overflow' count from a preceding unit being an input to the following unit.

Indication of the state (position) of the count can be provided by tapping points showing the state of that stage. A further requirement is a means of resetting all stages to off (0), to clear the circuit after making a count via a 'clear' signal — Fig. 11.1.

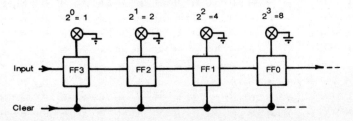

Fig. 11.1. Basic arrangement for a four-stage binary counter.

A four stage counter such as shown would then have a count capacity of $2^0 + 2^1 + 2^2 + 2^3 = 1 + 2 + 4 + 8 = 15$ decimal, although the actual number of combinations possible are $2^4 = 16$. The last pulse would produce 'overflow', i.e. returning all four stages to 0 and carrying a 1 on to a fifth stage, if present. The count capacity of such a stage is therefore $2^n - 1$, where n is the number of FF stages.

On the face of it, it would appear possible to use this 'spare' pulse to clear a $2^n - 1$ counter circuit. This is so, except that the process would be tedious. To clear after a count as many pulses would have to be applied to bring the count exactly to 2^n. Using a separate 'clear' signal, all stages can be returned to 0 with a single pulse.

Such a form of cascaded circuit is generally known as a *ripple counter* because the changes in outputs of the flip-flops ripple through the counter from input to output. A basic circuit and the corresponding waveform produced by a 4-stage ripple counter is shown in Fig. 11.2.

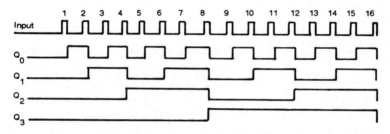

Fig. 11.2. Waveform signals in a ripple counter.

In practice, unless all the flip-flops change state simultaneously the waveforms may be spiked instead of square wave. It may therefore be necessary to treat the outputs in such a way that the counter is read only after these signals are stabilized.

The other main limitation of the ripple counter is that ripple-through delays are cumulative and where many stages are involved operating speed can be relatively slow. Such delays can be eliminated in a *synchronous* counter.

Reversible Counter

A *reversible counter* is one designed to count either forwards or backwards and is also known as an *up-down counter*. This is readily accomplished by using the Q output of the flip-flops for forward counting and the $\bar{Q}$ outputs for backward counting. The *direction* of counting is then determined by an up/down control signal X (e.g. X = 1 for up, X = 0 for down) applied to logic gates between the stages — Fig. 11.3.

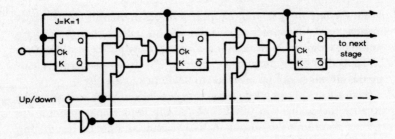

Fig. 11.3. Basic circuit for an up-down counter.

Decade Counters

It is often desirable to have the counter circuit count to base 10 instead of 2, i.e. in *decimal* rather than binary numbers. It is readily possible to utilize a ripple counter in this way, starting with the necessity of providing 10 combinations to cover a count of decimal 10. Again the least number of flip-flop stages (bits) required to do this will be four (i.e. giving $2^4 = 16$ possible combinations; $2^3 = 8$ would not be enough; and $2^5 = 32$ — far more than needed).

The basic circuit is then shown in Fig. 11.4. The principle involved is that at a count of 10 (binary 1010), all binaries are reset to zero via

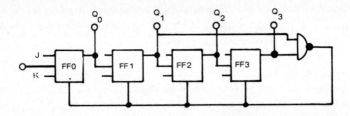

Fig. 11.4. Basic circuit for a decade (decimal) counter.

a feedback line containing a NAND gate, the output from which feeds all *clear* inputs in parallel. At a count of 10, output states are:

$$Q_0 = 0 \quad Q_1 = 1 \quad Q_2 = 0 \quad Q_3 = 0$$

Inputs to the NAND gate are thus Q1 and Q3. After the tenth pulse Q1 and Q3 both go to 1, the output of the NAND gate goes to 0, and FF0 and FF2 are reset to 0. Q1 and Q3 similarly return to 0 after a short delay.

This *propagation delay* can be troublesome unless eliminated, so the feedback line normally incorporates a *latching circuit* to memorize and hold the output of the NAND gate until all flip-flops clear.

To extend decimal counting beyond 10 it is only necessary to add further 4-bit counters in cascade, e.g.

to count to $100(10^2)$, 2 decade counters in cascade
$1000(10^3)$, 3 decade counters in cascade
$10000(10^4)$, 4 decade counters in cascade
and so on

Divide-by-N Counter

Exactly the same principle as used in the decade counter applies when designing a counter to count to any base N. The number of flip-flops required (n) is the smallest number for which $2^n > N$. Feedback via a NAND gate is then introduced to reset all binaries at the count of N, with each input to the NAND gate being an output from those flip-flops in state 1 at the count of N.

For example, a divide-by-5 counter would need three flip-flops. At the count of N = 5 their outputs would be:

$$Q_0 = 1 \quad Q_1 = 0 \quad Q_2 = 1$$

Hence Q_0 and Q_2 are the inputs to the NAND gate. Similarly, ripple-counter circuits can be designed to count in any BCD code required.

Some divide-by-N counters are *programmable*. That is to say they are designed to accommodate a number of different N values, selectable at will. Basically this only involves having a suitable number of flip-flops to start with and selecting the N setting by connection (or switching) the appropriate flip-flop outputs to the NAND gate inputs.

Synchronous Counters

In a *synchronous counter* circuit all flip-flops are clocked simultaneously by the input pulses. Speed of working is thus limited only by the delay time of any one flip-flop plus the propagation time of the control gate involved. In general terms this usually makes them about twice as fast as ripple-counters using similar components. There is also an absence of spikes in the output.

A typical basic circuit using T-type flip-flops is shown in Fig. 11.5. The requirement is that if T = 0 there is no change of state when the

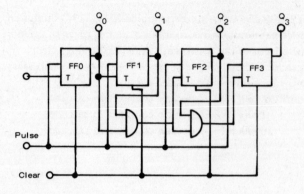

Fig. 11.5. Synchronous counter using T-type flip-flops.

binary is clocked; and if $T = 1$ the flip-flop output is complemented
with each pulse. In terms of T logic, this means:

$$T_0 = 1 \quad T_1 = Q_0 \quad T_2 = T_1.Q_1 \quad T3_2 = T_2.Q_2$$

This logic is performed by the AND gates.

A critical factor is the *minimum* time between pulses (T min) as this
governs the maximum signal pulse frequency which can be applied.
This is given by:

$$T \text{ min} = T_F + (n - 2)T_G$$

where T_F is the propagation time of one flip-flop

T_G is the propagation time of one AND gate

n is the number of AND gates

Maximum signal points frequency is then equivalent to $1/T$ min.

Speed of operation can be improved by parallel rather than series
working of the control gates, using a multiple-input AND gate taking
inputs from every preceding flip-flop. This does, however, have the
disadvantage of needing a large fan-in and fan-out, with heavier circuit loading. Nevertheless parallel working is widely used, particularly for synchronous forward-backward counters and decade
counters.

Synchronous Reversible Counter

A typical synchronous reversible (up-down) counter is shown in
Fig. 11.6. Again control gates are interposed between the flip-flops
but here they perform both up-down logic and (parallel) carry logic,
simplying the circuitry to some extent.

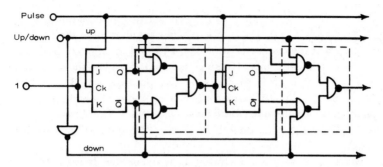

Fig. 11.6. Synchronous up/down counter.

Synchronous Divide-by-N Counters

Design of synchronous counter circuits for decade counters or divide-by-N working can be extremely tedious, but can be simplified by the use of Karnaugh maps. Numerous examples are, however, available in IC form and would normally be used in circuitry rather than start-from-scratch circuits. It is then only necessary to know the IC circuit characteristics and working parameters, and lead identification.

Johnson Counter (Twisted Ring Counter)

The circuit shown in Fig. 11.7 comprises five flip-flops connected with feedback from output to input thus resulting in a continuous loop or ring being formed. Because the ring is crossed over or 'twisted' at the input, it is known as a *twisted ring counter*. Alternatively, because it generates a Johnson code it is also called a *Johnson counter*.

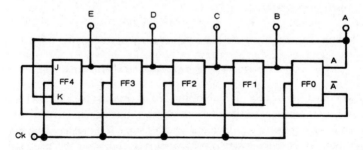

Fig. 11.7. Johnson (Twisted Ring) counter.

The working principle is as follows. Starting with all outputs zero ($A = 0$, $B = 0$, $C = 0$, $D = 0$, $E = 0$), after the first pulse the feedback loop applies the complement of A to FF4 and '1' appears at E. Successive pulses shift this '1' along the counter and at the same time insert '1's into ER4, so that after five pulses $A = 1$, $B = 1$, $C = 1$, $D = 1$, $E = 1$. The sixth pulse shifts $\bar{A}$ (i.e. 0) into FE4 and succeeding pulses similarly up to the ninth pulse when $A = 1$, $B = 0$, $C = 0$, $D = 0$, $E = 0$. The tenth pulse then shifts '0' into FE4 and all inputs are zero again.

In effect this circuit is a 1 to 10 (decimal) counter. In point of fact it has $2^5 = 32$ possible combinations, or a capacity to generate three different coded sequences of 10 decimal sequences.

IC *Binary Counter*

An example of an IC (HBC/HBF 4024A) providing a complete binary ripple through counter circuit is shown in Fig. 11.8. This is a

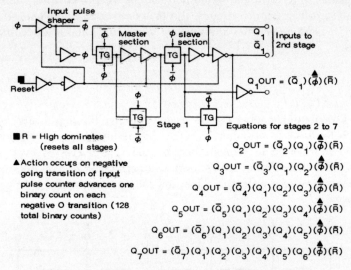

Fig. 11.8. IC binary ripple-through counter (HBC/HBF 4024A).

14-pin device available in dual in-line or flat ceramic package form. It provides an input pulse shaping circuit, reset line driven circuitry and seven binary counter stages. Each counter stage is a master-slave flip-flop and the counter stage is advanced one count on the negative-going transition of each input pulse. Typical speed of operation is

7 MHz at $V_{DD} - V_{SS} = 10$ volts. Total power dissipation is of the order of 200 milliwatts. The counter is reset to 0 by a HIGH level signal on the reset input. Functional circuitry is shown in Fig. 11.9.

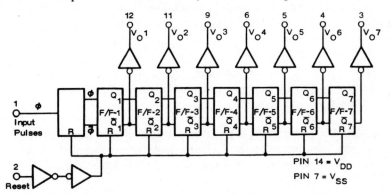

Fig. 11.9. Functional circuit of counter shown in Fig. 11.8.

Synchronous Counter

Fig. 11.10 shows a synchronous up/down counter which this time is a 16-pin package (HBC/HBF 4018A). It consists of 5-Johnson-

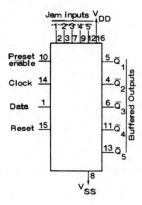

Fig. 11.10. IC Synchronous up/down counter (HBC/HBF 4018A).

counter stages, buffered $\bar{Q}$ outputs from each stage, and counter preset control gating. 'Clock', 'Reset', 'Data', 'Preset Enable' and 5 individual 'Jam' inputs are provided. Divide-by-10, 8, 6, 4, or 2 counter configurations can be implemented by feeding the $\bar{Q}5$, $\bar{Q}4$,

$\bar{Q}3$, $\bar{Q}2$, $\bar{Q}1$ signals, respectively, back to the Data input. Divide-by-9, 7, 5, or 3 counter configurations can be implemented by the use of a HBC/HBF 4011A gate package to properly gate the feedback connection to the Data input. Divide-by-functions greater than 10 can be achieved by use of multiple HBC/HBF 4018A units. The counter is advanced one count at the positive clock-signal transition. A Reset signal clears the counter to an 'all-zero' condition. A Preset-Enable signal allows information on the Jam inputs to preset the counter. Anti-lock gating is provided to assure the proper counting sequence.

Fig. 11.11 shows the logic elements and interval connections involved. This device has a maximum speed of working of 5 MHz at $V_{DD} - V_{SS} = 10$ volts. Total power dissipation is 200 milliwatts.

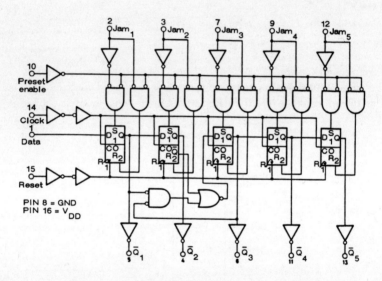

Fig. 11.11. Logic circuit of counter shown in Fig. 11.10.

CHAPTER 12

Converters and Registers

In a digital circuit data is represented by a series of digits, any change taking place in discrete steps. In many applications it is desirable to be able to present this data in the form of a continuous steady voltage or circuit which then varies smoothly with any change of state (i.e. an *analogue* signal infinitely variable between two limits). Systems for providing this are known as digital-to-analogue or *D/A converters*.

A basic form of 4-bit D/A converter is shown in Fig. 12.1 using a simple *weighted resistor network*. Input to each resistor, is via a digital switch (S0, S1, etc). When any switch is 'closed', i.e. equivalent to an

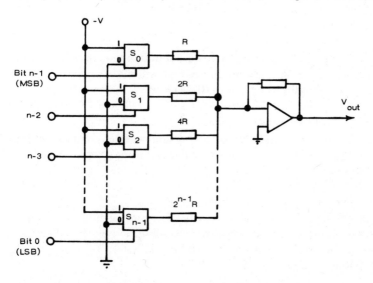

Fig. 12.1. Weighted resistor network digital-to-analogue converter.

input signal of 1, a constant reference voltage (V_R) is applied through the corresponding resistor. Resistor values are chosen so that the signal outputs in each line have 'weighted' values in a binary manner, i.e. 1, 2, 4, 8. Then a 1 at input S_0 gives an output of weighted value 1; a 1 at input S_1 an output of weighted value 2; a 1 at input S_2, an output of weighted value 4; a 1 at input V_3, an output of weighted value 8; and so on.

Put another way, since the same (constant) reference voltage (V) is applied to each line when the input to that line is 1, resistor values must be chosen so that:

line output voltage from S3 is twice that in line from S2
line output voltage from S2 is twice that in line from S1
line output voltage from S1 is twice that in line from S0.

This effectively gives 'weights' of 8, 4, 2 and 1 to the first four output lines, and so on. The total output voltage resulting from all lines is then fed to an operational amplifier (op-amp) to present the final output required as a current (the op-amp working as a voltage-to-current converter).

As an example, suppose the digital value is 1010 (decimal 10). Corresponding inputs will be:

$$\text{to } S3 = 1$$
$$\text{to } S2 = 0$$
$$\text{to } S1 = 1$$
$$\text{to } S0 = 0$$

If any input is 0 there will be no output in that line (the digital switch will remain 'open'). Output in this case will therefore be:

$$(1 \times 8) + (0 \times 4) + (1 \times 2) + (0 \times 1) = 10$$

That is to say the '1' inputs at S3 and S1 will give a final output of value 10 (the decimal equivalent).

The same principle can be extended to cover any number of bits. Thus for an N-bit D/A converter the following general relationship applies:

$$V \text{ out} = V_R \left(B_{n-1} 2^{-1} + B_{n-2} 2^{-2} + B_{n-3} 2^{-3} \dots B_0 2^{-n} \right)$$

where B_n represents the binary word.

This defines the *weighting* necessary. The most significant bit (B_{n-1}) has a weight of $V_R/2$, down to the least significant bit (B_0) which has a weight of $V_R/2^n$. Thus with a 6-bit converter, for example, the equation becomes:

$$V \text{ out} = \frac{VR}{64} \left(32n_5 + 16n_4 + 8n_3 + 4n_2 + 2n_1 + n_0 \right)$$

The basic disadvantages of such a circuit are: it demands stable, close tolerance resistors with values extending over a wide range; output resistance can be quite high; and the output signal is not a convenient multiple of the digital input value. Other circuits are therefore normally preferred in practice, e.g. a *serial* converter which works as an *integrator,* or a *ladder type* circuit.

The *ladder-type D/A converter* is more complex in that it requires twice the number of resistors to handle the same number of bits, but these only need to be of two values R and 2R. Actual resistor values are not so important as the correct 1:2 *ratio* values.

A basic circuit of this type is shown in Fig 12.2. Here the necessary 'weighting' of signals is achieved by *current splitting.* At the top of any 'ladder' the current will split equally right and left yielding weightings corresponding to $V_R/2$, $V_R/4$, $V_R/8$...down to $V_R/2^n$.

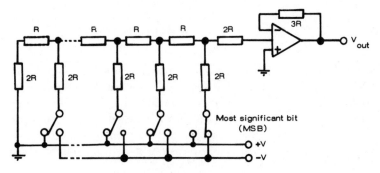

Fig. 12.2. Ladder-type D/A converter.

Analogue-to-Digital Converters

An *analogue-to-digital* (A/D) *converter* converts infinitely variable analogue data signals into digital form. There are many forms of such devices but the main types are voltage-to-frequency converters, pulse counters, and integrating converters.

Voltage-to-frequency converters are based on a voltage controlled oscillator where the output is applied to a counter for a period of time controlled by a clock pulse generator. Since this output frequency is proportional to input voltage the counter can be calibrated to read out the digital equivalent to the analogue input.

A basic example of a counter-type circuit is shown in Fig. 12.3. When an analogue signal (V_S) is applied to the comparator there is an output which 'opens' the gate, allowing clock pulses to be applied to the binary counter. The 'count' continues until the feedback signal (V_d) from the D/A converter becomes equal to V_S, when the comparator output falls to zero and the count is 'frozen' in the binary

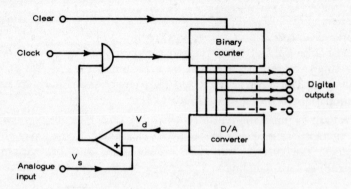

Fig. 12.3. Counter type analogue-to-digital converter.

counter and displayed or read-out. In other words the count proceeds one step at a time until final balance is reached. For example, to establish a count of 9.9 in 0.1 steps would involve 99 pulses passing through the gate before final balance is reached; or 999 pulses to count up to 99.9 with the same interval; and so on. The speed of conversion thus depends both on the pulse rate and the method by which final balance is obtained.

A more rapid method of conversion is possible using *successive approximations*. Here the first clock pulse sets the counter to one-half of the maximum output. The next pulse then sets the counter to one-half of a half in a 'plus' or 'minus' manner, i.e. 'plus' if V_S is greater than V_d' or 'minus' if V_d is greater than V_S; and so on with following pulses. This enables the final balance to be reached more quickly.

Shift Registers

A digital memory device has a 1-bit capacity so to store or *register* an N-bit word, N memory units (flip-flops) are required. It is then necessary to cascade the flip-flops output-to-output to feed input data into the system serially. It is this facility to 'shift' the data along the circuit that gives such a device the name *shift register*.

A basic circuit for performing this function is shown in Fig. 12.4. Each flip-flop is a master-slave type, the stage used to store the most significant bit (MSB) having S and R terminals connected together via an inverter to turn it into a D-type latch. Starting with all outputs clear ($Q_0 = 0$, $0_1 = 0$, etc) Cr is set to 1 and Pr held at 1 by keeping preset enable at 0.

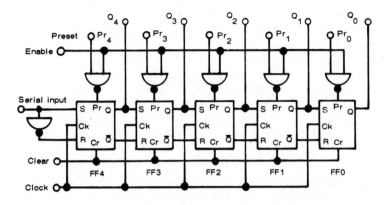

Fig. 12.4. Basic circuit for a 5-bit shift register.

Clock pulses are now applied. The first pulse (corresponding to the least significant bit) enters FF4 which latches, changing C_k from 0 to 1. Output Q4 is now at 1 with all other outputs at 0.

Each succeeding pulse then shifts the preceding pulse(s) to the right to make room for the incoming digit until after five pulses (or N pulses in an N-bit register), the full input word has been taken into the register. At that point the input pulses must stop. This sequence of operations can be seen from the following diagram, taking as an example 10110 as the 5-bit word fed into a 5-bit shift register.

clock pulse	word bit	MSB Q4	Q3	Q2	Q1	LSB Q0
1	1 →	1	0	0	0	0
2	0 →	0	1	0	0	0
3	1 →	1	0	1	0	0
4	1 →	1	1	0	1	0
5	0 →	0	1	1	0	1

Such a shift register accepts input *serially* and gives *parallel* output, and so is properly described as a *series-in, parallel-out register*. Other modes of working are possible. e.g.

    (i)   series-in, series-out
    (ii) parallel-in, parallel-out
    (iii) parallel-in, series-out

IC Shift Register

IC shift registers are produced in varying lengths and can be programmed to any number of bits between 1 and the maximum provided. An example is the (Mullard) HEF4557B 1-to 64-bit variable length shift register available as a flat 16-pin package with LSI — Fig. 12.5. The number of bits selected is equal to the sum of the subscripts

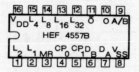

Fig. 12.5. IC 1-to 64-bit shift register (HEF 4557B).

of the enabled length control inputs (L1, L2, L4, L8, L16 and L32) plus 1, giving a maximum of 64. Serial data may be selected from D_A or D_B data inputs with the A/$\bar{\text{B}}$ select input. This feature is useful for recirculation purposes. Information on D_A or D_B is shifted into the first register position and all the data in the register is shifted one position to the right on the LOW to HIGH transition of CP_0 while $\overline{CP_1}$ is LOW; or on the HIGH to LOW transition of $\overline{CP_1}$ while CP_0 is HIGH.

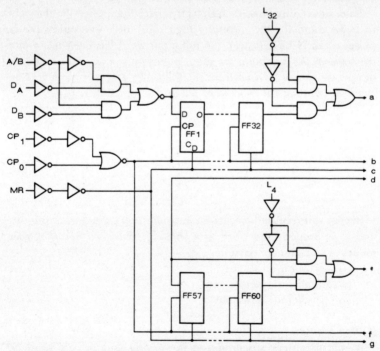

Fig. 12.6. Logic diagram for IC shift register of Fig. 12.5.

When HIGH, master reset (MR) resets the whole register asynchronously ($0 = $ LOW; $\overline{0} = $ HIGH) and independent of the other inputs. The complete logic diagram is show in Fig. 12.6.

This device can work on any voltage from 5-15v, drawing a quiescent current of 50-200μA. Propagation delay is of the order of 240-60 ns, depending on voltage, and maximum clock pulse frequency 5 MHz with 5-volt supply up to 20 MHz with 15-volt supply.

Another example of the logic provided by an IC shift register circuit is shown in Fig. 12.7. In effect this is a serial-to-parallel converter. In-

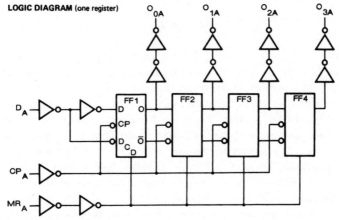

Fig. 12.7. IC series-to-parallel converter shift register (HEF 4058B).

formation preset on the data input D is shifted to the first register position and all the data in the register is shifted one position to the right by the clock pulse. The four outputs O_{0A}, 0_{1A}, 0_{2A}, 0_{3A}A are fully buffered. A HIGH (1) signal on the asynchronous master reset input MR clears the register and returns 0_0 to 0_3 to LOW (0), irrespective of the clock input and the serial data input (D_A).

This particular IC package (Signetics HEF 4015B) actually contains two such systems in a 16-pin flat package — Fig. 12.8.

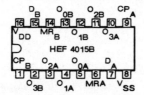

Fig. 12.8. IC package providing two shift registers.

The high component density which can be achieved with COSMOS integrated circuitry is well illustrated by the second 'packaged' register shown in Fig. 12.9. This is a 64-stage static shift register in

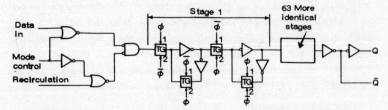

Fig. 12.9. CMOS 64-stage static shift register.

which each stage is a D-type master-slave flip-flop. The logic level present at the data input is transferred into the first stage and shifted one stage at each positive-going clock transition.

Information can be permanently shared with the clock line in either the LOW or HIGH state. There is also a mode input control which allows operation in the recirculating mode when in the HIGH state. Register packages can be cascaded and clock lines driven directly for fast working; or alternatively a delayed clock output is provided allowing reduced clock drive fan-out and transition time when cascaded.

The whole circuitry is contained in a 16-pin dual-in-line or ceramic flat package — Fig. 12.10.

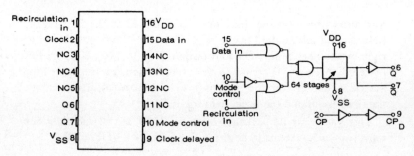

Fig. 12.10. CMOS HBC 4031A IC with logic diagram.

Dynamic MOS Shift Register

A basic circuit for a *dynamic* MOS *shift register* stage is shown in Fig. 12.11. It will be seen that it employs two separate clock inputs, i.e. is

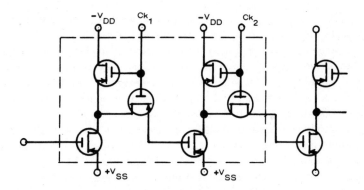

Fig. 12.11. Dynamic MOS shift register.

a two-phase MOS system, each stage incorporating SIX MOSFETs. Specifically these provide two NAND gates in cascade, each clock pulse shifting and inverting a bit through that stage.

Minimum and maximum clock rates are applicable — on minimum rate to retain gate capacitance, and a maximum rate limited by the response rate of the circuit.

It is a general feature of most IC dynamic MOS shift registers that both input and output are compatible with TTL integrated circuits.

The dynamic shift register shown in Fig. 12.12 provides 200 stages in a similar size package. It has provision for either single-on two-phase clock input signals. Single-phase-clocked operation is intended for low-power, low clock-line capacitance requirements. Single-phase

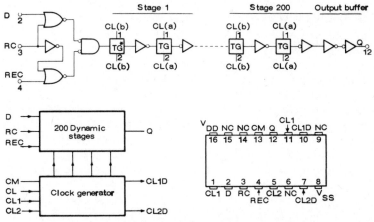

Fig. 12.12. IC shift register (HBC/HBE 4062A) with 200 dynamic stages.

clocking is specified for medium-speed operation (<1 MHz) at supply voltages up to 10v. Clock input capacitance is extremely low (<5 pF), and clock rise and fall times are non-critical. The clock-mode signal (CM) must be low for single-phase operation. Two-phase clock-input signals may be used for high-speed operation (up to 5 MHz) or to further reduce clock rise and fall time requirements at low speeds. Two-phase operation is specified for supply voltages up to 15v. Clock input capacitance is only 50 pF/phase. The clock-mode signal (CM) must be high for two-phase operation. The single-phase-clock input has an internal pull-down device which is activated when CM is high and may be left unconnected in two-phase operation. The logic level present at the data input is transferred into the first stage and shifted one stage at each positive-going clock transition for single-phase operation, and at the positive-going transition of CL1 for two-phase operation.

CHAPTER 13

Arithmetic Logic Units

Arithmetic logic units (ALU's) are specially designed for use in computers where they can be presented in highly compact form using LSI. Typically these provide 4-bit arithmetic operations with up to 16 instruction capability, e.g. addition, subtraction, up-down counting, shift, and logic operations. An arithmetical system of virtually any size can then be constructed by wiring together a number of ALU's.

In view of the significance of ALU's in computer technology, a detailed description is given of one particular 4-bit unit available in a 28-pin IC package only 35mm long and 15mm wide (Fig. 13.1), shown also in block diagram form in Fig. 13.2.

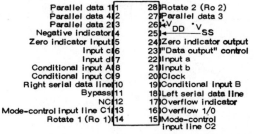

Parallel data 1 — 1	28 — Rotate 2 (Ro 2)
Parallel data 4 — 2	27 — Parallel data 3
Parallel data 2 — 3	26 — V_{DD} V_{SS}
Negative indicator — 4	25 —
Zero indicator Input — 5	24 — Zero indicator output
Input c — 6	23 — "Data output" control
Input d — 7	22 — Input a
Conditional input A — 8	21 — Input b
Conditional input C — 9	20 — Clock
Right serial data line — 10	19 — Conditional input B
Bypass — 11	18 — Left serial data line
NC — 12	17 — Overflow indicator
Mode-control input line C1 — 13	16 — Overflow 1/0
Rotate 1 (Ro 1) — 14	15 — Mode-control input line C2

Fig. 13.1. 4-bit arithmetic logic unit IC (HBC/HBF 4057A).

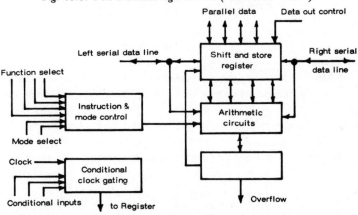

Fig. 13.2. Block diagram of 4-bit ALU.

The HBC/HBF 4057A arithmetic logic unit operates in one of four possible modes. These modes control the transfer of information, either serial data or arithmetic operation carries, to and from the serial-data lines. Fig. 13.3 shows the manner in which the four modes control the data on the serial-data lines.

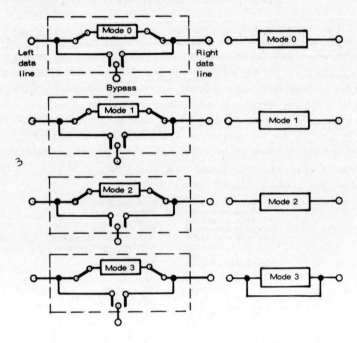

Fig. 13.3. Four modes of operation of ALU.

In Mode 0, data can enter or leave from either the left or right serial-data line.

In Mode 1, data can enter or leave only on the left serial-data line.

In Mode 2, data can enter or leave only on the right serial-data line.

In Mode 3, serial data can neither enter nor leave the register, regardless of the nature of the operation.

Furthermore, the register is by-passed electrically, i.e., there is an electrical bi-directional path between the right and left serial data terminals.

The two input lines labelled C1 and C2 in the terminal assignment diagram define one of four possible modes shown in Table 1.

Table 1, Mode Definition

C2	C1	Mode
0	0	0
0	1	1
1	0	2
1	1	3

Through the use of mode control, individual arithmetic arrays can be cascaded to form one large processor, or many processors of various lengths. Examples of how one 'hard-wired' combination of three ALU's can form (a) a 12-bit parallel processor, (b) one 8-bit and one 4-bit parallel processor, or (c) three 4-bit parallel processors, merely by changes in the modes of each ALU are shown in Fig. 13.4.

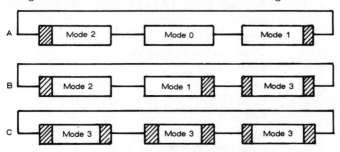

Fig. 13.4. Combination of three ALU's.

Data-flow interruptions are shown by shaded areas. With these three ALU's and the four available modes, 61 more system combinations can be formed. If 4 ALU's are used 4^4 combinations (256) are possible. fig. 13.5 shows a diagram of 4 HBC/HBF 4057A's interconnected to form a 16-bit parallel processor.

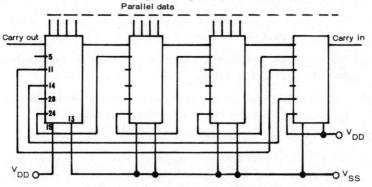

Fig. 13.5. Combination of four ALU's.

Instruction repertoire

Four encoded lines are used to represent 16 instructions.

Encoded instructions are as follows:

a	b	c	d	
0	0	0	0	NO-OP (Operational inhibit)
0	0	0	1	AND
0	0	1	0	Count down
0	0	1	1	Count up
0	1	0	0	Subtract stored number from zero (SMZ)
0	1	0	1	Subtract from parallel data lines (SM) (stored number from parallel data lines)
0	1	1	0	Add (AD)
0	1	1	1	Subtract (SUB) (Parallel data lines from stored number)
1	0	0	0	Set to all ones (SET)
1	0	0	1	Clear to all zeros (CLEAR)
1	0	1	0	Exclusive-OR
1	0	1	1	OR
1	1	0	0	Input Data (from parallel data lines)
1	1	0	1	Left shift
1	1	1	0	Right shift
1	1	1	1	Rotate (cycle) right

All instructions are executed on the positive edge of the clock.

Parallel commands

(a) CLEAR — sets register to zero.

(b) SET — sets register to all ones.

(c) OR — processes contents of register with value on parallel-data lines in a logical OR function.

(d) AND — processes contents of register with value on parallel-data lines in a logical AND function.

(e) Exclusive-OR — processes contents of register with data on parallel-data lines in a logical Exclusive-OR function.

(f) IN — loads data on parallel-data lines into register.

(g) DATA OUT CONTROL — unloads contents of register and overflow flip-flop on to parallel data lines and overflow I/O independent of all other controls.

(h) SUB:

In Mode 0, adds to the contents of the register the one's complement of the data on the parallel-data lines. Carries can enter on the right serial data line and can leave on the left serial data line. The overflow indicator does not change state.

In Mode 1, add to the contents of the register the 2's complement of the data on the parallel-data line. Generated carries can leave on the left serial line. The CARRY IN is set to zero. The overflow indicator does not change state.

In Mode 2, same as Mode 0, except carries cannot leave on the right serial-data line. The absence or presence of an overflow is registered.

In Mode 3, same as Mode 1, except carries cannot leave on the left serial-data line. The absence or presence of an overflow is registered.

(i) COUNT UP:

In Mode 0, adds to the contents of the register the data on the right serial-data line and permits any resulting carry to leave on the left serial-data line. No data enters the parallel-data lines.

In Mode 1, internally adds a one to the contents of the register and permits any resulting carry to leave on the left serial-data line. No data enters or leaves the right serial-data line.

In Mode 2, adds to the contents of the register the data on the right serial-data line. No data enters or leaves the left serial-data line.

In Mode 3, internally adds a one to the contents of the register. No data enters or leaves the register on any serial-data or parallel-data line.

In all modes, with the DATA OUT control 'high' the count is presented on the parallel data lines (D1-D4).

(j) COUNT DOWN:

In Mode 0, subtracts a one (2's complement form) from the contents of the register and adds to this result the data on the right serial-data line and permits any resulting carry to leave on the left serial-data line. No data enters on the parallel-data lines.

In Mode 1, internally subtracts a one from the contents of the register and permits any resulting carry to leave on the left serial-data line. No data enters or leaves the right serial-data line.

In Mode 2, subtracts a one from the contents of the register and adds to this result the data on the right serial-data line. No data enters or leaves on the left serial-data line.

In Mode 3, internally subtracts a one from the contents of the register. No data enters or leaves on the serial-data lines.

In all modes, with the DATA OUT control 'HIGH' the count is presented on the parallel data ines (D1-D4).

(k) ADD (AD):

In Mode 0, adds the contents of the register to the data on the parallel-data lines and the right serial-data line. Any resulting carry can leave on the left serial-data line. The overflow indicator does not change state.

In Mode 1, adds the contents of the register to the data on the parallel-data lines and allows any resulting carry to leave on the left serial-data line. The right serial-data line is open-circuited. The overflow indicator does not change state. The CARRY-IN is set to zero.

In Mode 2, adds the content of the register to the data on the parallel data lines and the right serial data line. Any overflow sets the overflow indicator. The left serial data line is open-circuited. The absence or presence of an overflow is registered.

In Mode 3, adds contents of the register to the data on the parallel-data lines. Any resulting carry sets the overflow indicator. The two serial-data lines are open circuited. The absence or presence of an overflow is registered. The CARRY-IN is set to zero.

(l) SM — same operation as AD except the contents of the register are 2's complemented during addition in Mode 1 and Mode 3. In Mode 2, the contents of the register are 1's complemented and added to the data on the right serial-data lines. Overflows occurring in Mode 1 or Mode 0 do not alter the overflow indicator. The presence or absence of overflows is registered on the overflow indicator in Mode 2 or Mode 3.

(m) SMZ:

In Mode 0, 1's complement the contents of the register and add the data on the right serial-data line to the contents of the register. Any resulting carry can leave on the left serial-data line. The overflow indicator does not change state.

In Mode 1, 2's complement the contents of the register and permit any carry to leave on the left serial-data line. No data can enter the right serial-data line. The overflow indicator does not change state. The CARRY-IN is set to zero.

In Mode 2, 1's complement the contents of the register and add the data on the right serial-data line to the contents of the register. Carries cannot leave the left serial data line. The absence or presence of an overflow alters the overflow indicator.

In Mode 3, 2's complement the contents of the register. Serial data can neither enter the right serial-data line nor leave the left serial-data line. The overflow indicator is at zero. The CARRY-IN is set to zero.

(n) NO-OP — no operation takes place. The clock input is inhibited and the state of all registers and indicators remains unchanged.

Serial-shift operations

(a) ROTATE (cycle) RIGHT — This operation is internal. The contents of the register shift to the right, cyclic fashion with the leftmost stage accepting data from the right-most stage regardless of the mode. Data can leave the register serially on the right data line only while the register is in Mode 2 or Mode 0. Data can enter the left data line serially while the register is in Mode 1 or Mode 0. The Ro1 terminal of the 'Most Significant' HBC/HBF 4057A must be connected to the Ro2 terminal of the 'Least Significant' HBC/HBF 4057A. All other Ro1 and Ro2 terminals must be left floating. When only one HBC/HBF 4057A is used, Ro1 must be connected to Ro2.

(b) RIGHT SHIFT — The contents of the register shift to the right and serial operations are as follows:

In Mode 0, data can enter serially on the left data line, shift through the register, and leave on the right data line.

In Mode 1, data can enter serially on the left data line. The right data line effectively is open-circuited.

In Mode 2, data can leave serially on the right data line. The left data line effectively is open-circuited. Vacant spaces are filled with zeros.

In Mode 3, serial data can neither enter nor leave the register; however, the contents shift to the right and vacated places are filled with zeros.

In all modes, with the DATA OUT control 'HIGH' the data is presented on the parallel data line (D1-D4).

(c) LEFT SHIFT — The contents of the register shift to the left and serial operations are as follows:

In Mode 0, data can leave serially on the left data line. The right data line effectively is open-circuited. All vacant positions are filled with zeros.

In Mode 2, data can enter serially on the right data line. The left data line effectively is open-circuited.

In Mode 3, data can neither enter nor leave the register; however, the contents shift to the left, and vacated places are filled with zeros.

In all modes, with the DATA OUT control 'HIGH' the data is presented on the parallel data lines (D1-D4).

Because the 'DATA OUT' control instruction is independent of the other 16 instructions, care must be taken not to activate this control when data are to be loaded into the processor. This instruction should only be activated when the processor is executing a NO-OP, any SHIFT, SMZ, COUNT UP or DOWN, CLEAR or SET.

If a data line, serial or parallel, is used as an input and the logic state of that line is not defined (i.e. the line is an open circuit), then the result of any operation using that line is undefined.

Operational sequence for arithmetic add cycle

1. Apply IN Instruction and Word A on Parallel Data Lines (D1-D4).
2. Apply CLOCK to load Word A into the register.
3. Apply OP CODE Instruction and Word B on Data Lines.
4. Apply CLOCK to load resulting function of A and B into the register.
5. Apply 'DATA OUT' control to present result to Parallel Data Lines.

Zero detection

The condition of 'all zeros' is indicated by a '1' on the Zero IND terminal of the 'Most Significant' HBC/HBF 4057A.

Negative-number detection

The NEG IND terminal of the HBC/HBF 4057A is connected to the output of the flip-flop that is in the most significant bit position. A '1' on the NEG IND terminal indicates a negative number is in the register. This detection is independent of modes.

Complementing numbers

1. 1's complement of number in ALU register:
   (a) ALU must be in MODE 0 or MODE 2.
   (b) Zero on Rt. Data Line.
   (c) Execute an SMZ instruction.
2. 1's complement of number to be loaded into ALU register.
   (a) If zero indicator output is low, execute a CLEAR instruction, and make Rt. Data Line = 0.
   (b) ALU must be in MODE 0 or MODE 2.
   (c) Execute a SUB instruction.
3. 2's complement of number in ALU register.
   (a) ALU must be in MODE 1 or MODE 3.
   (b) Execute an SMZ instruction.
4. 2's complement of number to be loaded into ALU register.
   (a) If zero indicator output is low, execute a CLEAR instruction, and make Rt. Data Line = 0.
   (b) ALU must be in MODE 1 or MODE 3.
   (c) Execute a SUB instruction.

The following algorithms are given as a general guideline to demonstrate some of the capabilities of the HBC/HBF 4057A.

Multiplication of two N-bit numbers

Multiplication Algorithm
1. Clear ALU to Zero.
2. Store $A_s.B_s$ in External Flip-flop.
3. If $A_s = 1$, Complement Register 1.
4. If $B_s = 1$, Complement Register 2.
5. Load Register 2 into ALU.
6. Do Shift Left on ALU N Times (N = number of bits).

7. Do N Time:
   (a) If MSB of ALU = 1 (Negative Indicator = High). Then shift ALU left 1 bit; add Register 1 to ALU.
   (b) If MSB of ALU = 0 (Negative Indicator = Low). Then shift ALU left 1 bit.
8. If $A_s.B_s = 1$, then Complement ALU.
9. Answer in ALU.

Division Algorithm
1. Store $A_s.B_s$ in External Flip-flop.
2. If $A_s = 1$, complement ALU 1 and ALU 2.
3. If $B_s = 1$, complement Register A.
4. Check for Divisor = 0:
   (a) If Divisor = 0; stop, indicates division by 0.
  (b) If Divisor = 0; continue.
5. Apply SUB instruction to ALU 1 and Register A to ALU 1 data lines.
   (a) If $C_o = 0$ (Dividend <Divisor), Stop, indicates overflow.
   (b) If $C_o = 1$ (Dividend >Divisor), Continue.
6. Put a zero on RT, data line of ALU 2 and shift ALU 1 and ALU 2 left 1 bit.
7. Do 'N' times.
   (a) If $C_o = 1$, then clock ALU 1, and put a 1 on right data line of ALU 2.
   (b) If $C_o = 0$, then no clock, and put a 0 on right data line of ALU 2.
8. If sign Flip-flop = 1, complement ALU 2.
9. Answer in ALU 2.

Conditional operation

Inhibition of the clock pulse can be accomplished with a programmed NO-OP instruction or through conditional input terminals A, B and C. In a system of many HBC/HBF 4057A's, each HBC/HBF 4057A can be made control automatically its own operation or the operation of any other HBC/HBF 4057A in the system in conjunction with the Overflow, Zero, or Negative (Number) indicators. Table II, the conditional-inputs truth table, defines the interaction among A, B and C.

Table II. Conditional Inputs Truth Table

A	B	C	Operation Permitted
0	X	X	yes
1	0	0	yes
1	0	1	no
1	1	0	no
1	1	1	yes

X = 'don't know'

Two examples of how the conditional operation can be used are as follows:

1. For the Multiplication Algorithm

   A = 1, for step 7 (1)

   A = 0, for step 7 (2)

   B = 1

   C = negative indicator.

2. For the Division Algorithm

   A = 1, for step 7 (1)

   A = 0, for step 7 (2)

   B = 1

   C = C_o (left data line).

Overflow detection

The HBC/HBF 4057A is capable of detecting and indicating the presence of an arithmetic 2's-complement overflow. A 2's-complement overflow is defined as having occurred if the signs of the two initial words are the same and the sign of the result is different while performing a carry-generating instruction.

$$
\begin{array}{r}
0.011 \\
(+)\ \underline{0.110} \\
1.001
\end{array}
$$

For example:

Overflows can be detected and indicated only during operation in Mode 2 or Mode 3 and can occur for only three instructions (AD, SMZ, and SUB). If an overflow is detected and stored in the overflow flip-flop, any one of the five instructions AD, SMZ, SM, SUB, or IN can change the overflow indicator. When any of the three subtraction instructions is used, the sign bit of the data being subtracted is complemented and this value is used as one of the two initial signs to detect overflows. If an overflow has occurred, the final sign of the sum or difference is 1's complemented and stored in the most-significant-bit position of the register.

The overflow flip-flop is updated at the same time the new result is stored in the HBC/HBF 4057A. Whenever data on the parallel-data lines are loaded into the HBC/HBF 4057A, whatever is on the Overflow I/O line is loaded into the overflow flip-flop.

Also, whenever data are dumped on the parallel data lines from the HBC/HBF 4057A, the contents of the overflow flip-flop are dumped on the Overflow I/O line. Thus overflows may be stored elsewhere and then fed into the HBC/HBF 4057A at another time.

Appendix I

BINARY/DECIMAL EQUIVALENTS

Decimal	2^5	2^4	2^3	2^2	2^1	2^0
	(32)	(16)	(8)	(4)	(2)	(1)
0						0
1						1
2					1	0
3					1	1
4				1	0	0
5				1	0	1
6				1	1	0
7				1	1	1
8			1	0	0	0
9			1	0	0	1
10			1	0	1	0
11			1	0	1	1
12			1	1	0	0
13			1	1	0	1
14			1	1	1	0
15			1	1	1	1
16		1	0	0	0	0
17		1	0	0	0	1
18		1	0	0	1	0
19		1	0	0	1	1
20		1	0	1	0	0
21		1	0	1	0	1
22		1	0	1	1	0
23		1	0	1	1	1
24		1	1	0	0	0
25		1	1	0	0	1
26		1	1	0	1	0
27		1	1	0	1	1
28		1	1	1	0	0
29		1	1	1	0	1
30		1	1	1	1	0
31		1	1	1	1	1
32	1	0	0	0	0	0
33	1	0	0	0	0	1
34	1	0	0	0	1	0
35	1	0	0	0	1	1
36	1	0	0	1	0	0
37	1	0	0	1	0	1
38	1	0	0	1	1	0
39	1	0	0	1	1	1
40	1	0	1	0	0	0
41	1	0	1	0	0	1
42	1	0	1	0	1	0
43	1	0	1	0	1	1

Decimal	2^6 (64)	2^5 (32)	2^4 (16)	2^3 (8)	2^2 (4)	2^1 (2)	2^0 (1)
44		1	0	1	1	0	0
45		1	0	1	1	0	1
46		1	0	1	1	1	0
47		1	0	1	1	1	1
48		1	1	0	0	0	0
49		1	1	0	0	0	1
50		1	1	0	0	1	0
51		1	1	0	0	1	1
52		1	1	0	1	0	0
53		1	1	0	1	0	1
54		1	1	0	1	1	0
55		1	1	0	1	1	1
56		1	1	1	0	0	0
57		1	1	1	0	0	1
58		1	1	1	0	1	0
59		1	1	1	0	1	1
60		1	1	1	1	0	0
61		1	1	1	1	0	1
62		1	1	1	1	1	0
63		1	1	1	1	1	1
64	1	0	0	0	0	0	0
65	1	0	0	0	0	0	1
66	1	0	0	0	0	1	0
67	1	0	0	0	0	1	1
68	1	0	0	0	1	0	0
69	1	0	0	0	1	0	1
70	1	0	0	0	1	1	0
71	1	0	0	0	1	1	1
72	1	0	0	1	0	0	0
73	1	0	0	1	0	0	1
74	1	0	0	1	0	1	0
75	1	0	0	1	0	1	1
76	1	0	0	1	1	0	0
77	1	0	0	1	1	0	1
78	1	0	0	1	1	1	0
79	1	0	0	1	1	1	1
80	1	0	1	0	0	0	0
81	1	0	1	0	0	0	1
82	1	0	1	0	0	1	0
83	1	0	1	0	0	1	1
84	1	0	1	0	1	0	0
85	1	0	1	0	1	0	1
86	1	0	1	0	1	1	0
87	1	0	1	0	1	1	1
88	1	0	1	1	0	0	0
89	1	0	1	1	0	0	1
90	1	0	1	1	0	1	0
91	1	0	1	1	0	1	1

Decimal							
92	1	0	1	1	1	0	0
93	1	0	1	1	1	0	1
94	1	0	1	1	1	1	0
95	1	0	1	1	1	1	1
96	1	1	0	0	0	0	0
97	1	1	0	0	0	0	1
98	1	1	0	0	0	1	0
99	1	1	0	0	0	1	1
100	1	1	0	0	1	0	0
101	1	1	0	0	1	0	1
102	1	1	0	0	1	1	0
103	1	1	0	0	1	1	1
104	1	1	0	1	0	0	0
105	1	1	0	1	0	0	1
106	1	1	0	1	0	1	0
107	1	1	0	1	0	1	1
108	1	1	0	1	1	0	0
109	1	1	0	1	0	0	1

Decimal	2^7	2^6	2^5	2^4	2^3	2^2	2^1	2^0
	(128)	(64)	(32)	(16)	(8)	(4)	(2)	(1)
110		1	1	0	1	1	1	0
111		1	1	0	1	1	1	1
112		1	1	1	0	0	0	0
113		1	1	1	0	0	0	1
114		1	1	1	0	0	1	0
115		1	1	1	0	0	1	1
116		1	1	1	0	1	0	0
117		1	1	1	0	1	0	1
118		1	1	1	0	1	1	0
119		1	1	1	0	1	1	1
120		1	1	1	1	0	0	0
121		1	1	1	1	0	0	1
122		1	1	1	1	0	1	0
123		1	1	1	1	0	1	1
124		1	1	1	1	1	0	0
125		1	1	1	1	1	0	1
126		1	1	1	1	1	1	0
127		1	1	1	1	1	1	1
128	1	0	0	0	0	0	0	0
129	1	0	0	0	0	0	0	1
130	1	0	0	0	0	0	1	0
131	1	0	0	0	0	0	1	1
132	1	0	0	0	0	1	0	0
133	1	0	0	0	0	1	0	1
134	1	0	0	0	0	1	1	0
135	1	0	0	0	0	1	1	1
136	1	0	0	0	1	0	0	0
137	1	0	0	0	1	0	0	1
138	1	0	0	0	1	0	1	0

Decimal	2^7	2^6	2^5	2^4	2^3	2^2	2^1	2^0
	(128)	(64)	(32)	(16)	(8)	(4)	(2)	(1)
139	1	0	0	0	1	0	1	1
140	1	0	0	0	1	1	0	0
141	1	0	0	0	1	1	0	1
142	1	0	0	0	1	1	1	0
143	1	0	0	0	1	1	1	1
144	1	0	0	1	0	0	0	0
145	1	0	0	1	0	0	0	1
146	1	0	0	1	0	0	1	0
147	1	0	0	1	0	0	1	1
148	1	0	0	1	0	1	0	0
149	1	0	0	1	0	1	0	1
150	1	0	0	1	0	1	1	0
151	1	0	0	1	0	1	1	1
152	1	0	0	1	1	0	0	0
153	1	0	0	1	1	0	0	1
154	1	0	0	1	1	0	1	0
155	1	0	0	1	1	0	1	1
156	1	0	0	1	1	1	0	0
157	1	0	0	1	1	1	0	1
158	1	0	0	1	1	1	1	0
159	1	0	0	1	1	1	1	1
160	1	0	1	0	0	0	0	0
up to								
255	1	1	1	1	1	1	1	1
and so on								

Higher orders of binary number equivalents:

2^8	256	2^{30}	1073741824
2^9	512	2^{31}	2147483648
2^{10}	1024	2^{32}	4294967296
2^{11}	2048	2^{33}	8589934592
2^{12}	4096	2^{34}	17179869184
2^{13}	8192	2^{35}	34359738368
2^{14}	16384	2^{36}	68719476736
2^{15}	32768	2^{37}	137438953472
2^{16}	65536	2^{38}	274877906944
2^{17}	131072	2^{39}	549755813888
2^{18}	262144	2^{40}	1099511627776
2^{19}	524288	2^{41}	2199023255552
2^{20}	1048576	2^{42}	4398046511104
2^{21}	2097152	2^{43}	8796093022208
2^{22}	4194304	2^{44}	17592186044416
2^{23}	8388608	2^{45}	35184372088832
2^{24}	16777216	2^{46}	70368744177664
2^{25}	33554432	2^{47}	140737488355328
2^{26}	67108864	2^{48}	281474976710656
2^{27}	134217728	2^{49}	562949953421312
2^{28}	268435456	2^{50}	1125899906842624
2^{29}	536870912		

Appendix II

Truth Tables, Boolean Algebra and Minimizing

As an example, suppose the problem involves three inputs A, B and C and the logic to be provided is that there is an output with the following combinations of signals:

B *and* C *or* A *and* C *or* A *and* B

The corresponding Truth Table can be written:

	A	B	C	S
line 1	0	0	0	0
line 2	0	0	1	0
line 3	0	1	0	0
line 4	0	1	1	1
line 5	1	0	0	0
line 6	1	0	1	1
line 7	1	1	0	1
line 8	1	1	1	1

This describes all 'states', i.e. twenty four possible combinations, of which only four provide one output. (This is not so obvious from the original statement, where it may appear that only *three* states will given an output).

The solution for S can also be written in the form of a Boolean equation (remembering that . means AND and + means OR):

$$S = 1 = \bar{A}.B.C. + A.\bar{B}.C + A.B.\bar{C} + A.B.C$$

Again this indicates four states giving an output (S = 1).

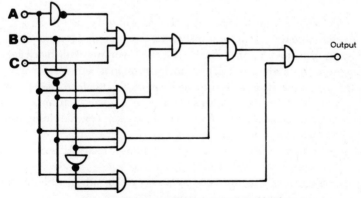

Fig. A1. Unminimized logic element combination.

These states can be provided by covering all states as laid down by the original logic statement, reminded by the truth table, and/or the Boolean equation that they are *four* possible states involved for S = 1. Logic elements arranged in the combination shown in Fig. A1 would then cover *all* states with *unminimized* combinations. A second look with a view to *minimizing* the actual combinations required can be very worthwhile.

The Boolean equation simplifies to:
    S = 1 = B.C. + A.C + A.B
which is merely the original statement expressed in Boolean algebra. The same follows from a study of the truth table. We only *need* the states established by the fourth, sixth and seventh lines.

The *minimized* circuit is then very much simpler, reducing the number of logic elements *actually* required from ten to five — Fig. A2.

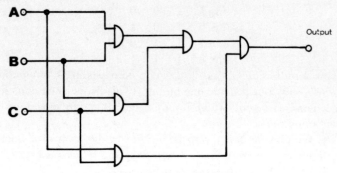

Fig A2. Minimized logic element circuit.

Minimizing is thus a very important part of logic circuit design. It can eliminate redundant or unnecessary components. It is not always easy to spot how for this can be achieved working with block logic diagrams, but reducing the Boolean equation to its simplest form provides a positive answer — provided you do the Boolean algebra correctly. It is not so easy with *combination circuits* where minimizing is best done with the aid of a *Karnaugh-Veitch map*. The full design procedure then involves:
    (i) Putting down all input combinations which will provide an output.
    (ii) Construct a truth table as a check that all possible combinations have been considered.

(iii) Derive the Boolean equation which will provide an output.
(iv) Construct a Karnaugh-Veitch map for all participating variables.
(v) Use this map to *minimize* equation (iii)
(vi) Draw up the circuit from this minimized equation.

Minterms and Maxterms

Truth tables and (Boolean) algebra equations are closely related. In most digital circuit designs the starting point is the truth table from which the corresponding formula is derived.

Example: truth table

A	B	S
0	0	0
0	1	1
1	0	1
1	1	0

(Note. This is the truth table for exclusive OR logic)

Corresponding formula $A\bar{B} + \bar{A}B = S$

This particular formula is a 'sum of products', or what is called the normal *minterm* form when referring to switching circuits.

A complementary formula can be devised for the same conditions (i.e. from the same truth table) by considering combinations which do *not* produce an output. This is called a *dual* equation and in this case would be:

$$\bar{A}\bar{B} + AB = \bar{S}$$

$$\text{by inversion } S = \overline{\bar{A}\bar{B} + AB}$$
$$= (A + B)(\bar{A} + \bar{B})$$

This is a 'product of sums' and this form of equation is called the normal *maxterm* form.

Specifically, then, the *minterm* form of an equation, being a *sum* (of products) can be solved by digital devices having an AND function. *Maxterm* forms of an equation, being a *product* (of sums) can be solved by OR devices.

The value of this is that a switching function requirement can be written in equations in either minterm for solution with AND devices, or maxterm form for solution with OR devices, and the two alternatives compared term for term.

Again minterms and maxterms can be directly related to a truth table. For example, possible minterms and maxterms covering three binary variables A, B and C are:

A	B	C	minterm	maxterm
0	0	0	$\bar{A}\bar{B}\bar{C}$	$\bar{A}+\bar{B}+\bar{C}$
0	0	1	$\bar{A}\bar{B}C$	$\bar{A}+\bar{B}+C$
0	1	0	$\bar{A}B\bar{C}$	$\bar{A}+B+\bar{C}$
0	1	1	$\bar{A}BC$	$\bar{A}+B+C$
1	0	0	$A\bar{B}\bar{C}$	$A+\bar{B}+\bar{C}$
1	0	1	$A\bar{B}C$	$A+\bar{B}+C$
1	1	0	$AB\bar{C}$	$A+B+\bar{C}$
1	1	1	ABC	$A+B+C$

Minterms and maxterms can also be devised directly from any functional expression f(A,B) where f is a Boolean function of the binary variables A and B.

Minterm form $f(A,B) = \bar{A}\bar{B}.f(0,0) + \bar{A}B.f(0,1) +$
$$AB.f(1,0) + AB.f(1,1)$$

Maxterm form $f(A,B) = (\bar{A} + B + f(1,1)).(\bar{A} + B + f(1,0).$
$$(A + \bar{B} + f(0,1).(A + B + f(0,0)$$

Karnaugh Maps

In a Karnaugh map every possible combination of the binary input variables is represented by a square (called a *cell*). The number of squares required will be equal to 2^n, where n is the number of variables to cover all possible combinations. There will be $2^2 = 4$ squares.

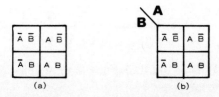

Fig. A3. Karnaugh maps for two binary variables.

Taking the simplest case of two variables A and B, the map will have four cells, with the four possible combinations allocated as in Fig. A3(a). Alternatively the signal values can also be indicated — Fig. A3(b). In each case information about the variable A is contained in the pair of cells in the right-hand *column* and information about the variable B in the *bottom pair* of cells.

Each cell is defined by the intersection of two variables, e.g. the top left-hand cell is defined by the intersection of A AND B, and thus obviously has a value of 0. Similarly the values for other cells can be determined and inserted, rendering the map in the form of Fig. A4.

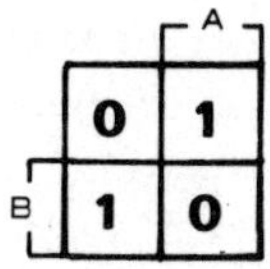

Fig. A4. Cells with binary values.

The sequence in which the variables are presented is not significant but what is important is that each variable defines one half of each of the total available cells and is linked with as half of the cells associated with each of the other variables. On this basis, maps for three and four variables are shown in Figs. A5 and A6 respectively.

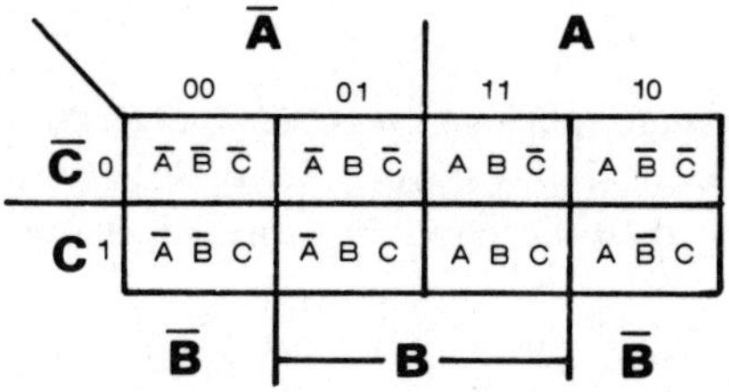

	$\bar{A}$		A	
	00	01	11	10
$\bar{C}$ 0	$\bar{A}\ B\ \bar{C}$	$\bar{A}\ B\ \bar{C}$	$A\ B\ \bar{C}$	$A\ \bar{B}\ \bar{C}$
C 1	$\bar{A}\ \bar{B}\ C$	$\bar{A}\ B\ C$	$A\ B\ C$	$A\ \bar{B}\ C$
	$\bar{B}$	$-$B$-$		$\bar{B}$

Fig. A5. Karnaugh map for three variables.

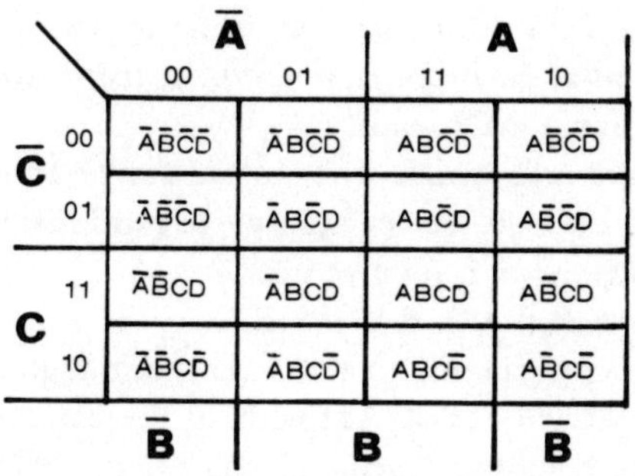

	$\bar{A}$		A	
	00	01	11	10
$\bar{C}$ 00	$\bar{A}\bar{B}\bar{C}\bar{D}$	$\bar{A}B\bar{C}\bar{D}$	$AB\bar{C}\bar{D}$	$A\bar{B}\bar{C}\bar{D}$
01	$\bar{A}\bar{B}\bar{C}D$	$\bar{A}B\bar{C}D$	$AB\bar{C}D$	$A\bar{B}\bar{C}D$
C 11	$\bar{A}\bar{B}CD$	$\bar{A}BCD$	$ABCD$	$A\bar{B}CD$
10	$\bar{A}\bar{B}C\bar{D}$	$\bar{A}BC\bar{D}$	$ABC\bar{D}$	$A\bar{B}C\bar{D}$
	$\bar{B}$	B		$\bar{B}$

Fig. A6. Karnaugh map for four variables.

Simple Operation

An example of the use of Karnaugh maps is to determine the results of AND or OR gating of two or more functions which may themselves involve fairly complex logic. For simplicity, assume that both A and B have two variables and a quick check is required on the results of combining A and B as an AND function.

A and B are both drawn as separate maps, with respective cell values and annotated by an AND sign (.) — Fig. A7. It is then readily possible to plot the resulting 'A.B' map.

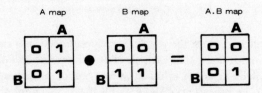

Fig. A7. A.B map derived from A and B maps.

Equally, Karnaugh maps can be used to determine the *inverse* of a function simply by changing 0's to 1's and 1's to 0's in the individual maps, remembering at the same time this will change AND to OR (or vice versa).

Minimization Techniques

Probably the most useful application of Karnaugh maps is to minimize the number of logic elements necessary to provide a solution to the problem displayed on the map. This involves grouping together adjacent cells on the map with the object of arriving at the simplest statement of the original equation or truth table by graphical rather than mathematical means.

Adjacent cells are cells which differ in only one variable in the AND terms describing the cells. As an example of minimization technique, suppose the logic equation involved is:

$$f = \bar{A}.B.\bar{C} + \bar{A}.B.C + A.B.C + A.B.\bar{C}$$

This, in fact, corresponds to the basic Karnaugh map configuration for three variables (Fig. A5) without the combinations $\bar{A}.\bar{B}.\bar{C}$ and $\bar{A}.\bar{B}.C$. The resulting Karnaugh map is therefore as shown in Fig. A8. The individual cells are designated (i) (ii) (iii), etc., for reference description only. They would not normally be so marked.

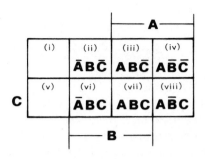

Fig. A8. 8-cell Karnaugh map for three variables.

Adjacent cells (ii) and (iii) differ in only one variable (A and $\bar{\text{A}}$), and can thus be grouped. Similar adjacent cells (vi) and (vii) differ only in one variable (A and $\bar{\text{A}}$ again) and can thus be grouped — Fig. A9(a). Having done this the whole map can be defined in the simpler expression:

$$f = B.\bar{C} + B.C$$

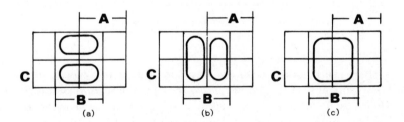

Fig. A9. Grouping of adjacent cells.

Similarly cells can be grouped in columns, as in Fig. A9(b). In this case the simplified equation becomes:

$$f = A.B + \bar{A}.B$$

Equally, in this example adjacent cells (ii), (iii), (vi) and (vii) *all* differ only in one variable, so these four can be grouped. This, in fact, reduces the equation to its simplest possible form — (Fig. A9(c)):

$$f = B$$

Appendix III

COMPUTER PROGRAMMING

A digital computer consists essentially of a *store unit* or memory for numbers and instructions; an *arithmetic unit* where arithmetical operations are performed on the numbers; a *control unit* for controlling and sequencing the operations correctly; an *input unit* to transfer data into the computer; and an *output unit* to transfer data out of the computer store. All these units are interconnected in the manner shown in the block diagram — Fig. A10.

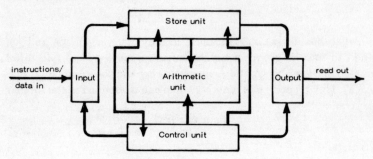

Fig. A10. Block diagram of computer.

Both instructions and stored data are in the form of computer *words* consisting of so-many *bits*. Typically the length of such words ranges from 18 to 40 bits. Stored information is located at a particular *address*. Instructions fed into the computer to derive answers are directed into an order and address, separated by a *modifier* bit — Fig. A11.

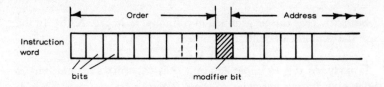

Fig. A11. Computer instruction word.

The order part of the instruction word must contain sufficient bits to cover all the orders within the capabilities of the computer to perform, specified as *machine code orders* and designated by a specific code. Similarly, the address which follows is also fed in in binary code. The purpose of the modifier bit between the two is to instruct the control unit to add the contents of the *modifier register* to the instruction before it is obeyed, these instructions being included in the machine code. The use of modifier bits in this way greatly increases the speed of working and efficiency of the computer.

Taking a computer capable of accommodating an 18-bit word as an example, the instruction word will have a capacity of $2^{18} = 4130$ orders. If 5-bits are allocated for the order, this will provide a machine code order length of $25 = 32$ separate orders, leaving $2^{12} = 4096$ separate locations available for addresses. (The other bit making a total of 2^{15} is the modifier bit).

Machine codes may be specified in numerical order, or mnemonics, the latter being generally easier to remember. An example of a machine code order covering 32 separate orders in numerical and mnemonic codes (from *Logical Design of Switching Circuits* by Douglas Lewin) is:

Order code	Mnemonic	Order
00	IN	Input character from paper tape to accumulator
01	OUT	Output character from accumulator to paper tape
02		
03		
04		
05		
06		
07		
10	ADD	Add contents of address specified in instruction to accumulator. Location specified is unchanged.
11	SUB	Subtract contents of address specified in instruction from accumulator, location specified is unchanged.
12	FET	Fetch contents of address specified in instructions to accumulator, location specified is unchanged.

13	STR	Store contents of accumulator in address specified in instruction, accumulator remains unchanged.
14	COM	Take 2's complement of accumulator contents
15	SHFL	Shift accumulator contents.
16	SHFR	Shift accumulator right N places (N specified by address digits).
17	COL	Collate accumulator with contents of the address specified in instruction, location unchanged.
20	ADDM	Add +1 to modifier register
21	FETM	Fetch contents of modifier to accumulator modifier unchanged.
22	STRM	Store contents of accumulator in modifier, accumulator unchanged.
23	LINK	Store contents of instruction register in address specified in instruction and jump to address specified +1.
24		
25		
26		
27		
30	JMP	Jump to address specified in instruction and take next instruction from there.
31	JMPN	Jump to address specified in instruction if accumulator negative, otherwise take next instruction in sequence.
32	JMPP	Jump to address specified in instruction if accumulator positive, otherwise take next instruction in sequence.
33	JMPO	Jump to address specified in instruction if accumulator zero, otherwise take next instruction in sequence.
34	STOP	Stop.
35	JMPO	Jump to address specified in instruction if modifier zero, otherwise take next instruction in sequence.

The store contains both *instructions* (e.g. add, subtract, multiply, etc.) and *data,* and different addresses. The object of the programme is to search out both from their respective addresses and bring them together to complete the programme.

Taking a simple example, suppose the problem is to add A to B. The programme would then be written as follows:

		Order	Address
(i)	(12)	FET	Address of A
(ii)	(10)	ADD	Address of B
(iii)	(13)	STR	–
(iv)	(34)	STOP	–

The corresponding working is:
  (i)   Fetch A from address to accumulator
  (ii) Fetch B from address and add in to accumulator
  (iii) Store A + B in accumulator
  (iv) Stop

The computer has now completed its programme and is ready to commence its next 'read' cycle, i.e. obey any further instruction relative to the problem. If there are no further instructions then the answer is read-out.

The same basic principle applies, regardless of the complexity of the programme. However there are simplifying techniques which can be used to speed processing, using data stored in *modifying registers* and called into the programme by a modifier bit. Repetition work may be conducted in a separate link by sub-routing at a particular stage in the programme and then feeding back to continue in the main programme.

BASIC — *and other Computer Languages*

Simplified forms of programming language have been developed for home computers (and others), making it possible to type in statements (computer commands) in ordinary language, conforming to a specific *listing* (written record) of the programme. The most widely used language for home computers is BASIC (and its variations) in which statements and/or questions and inputs (instructions and answers) are typed into the computer in logical order. The computer will then print out (i.e. display the complete text on a TV monitor) in numerical order, unless one instruction commands the computer to 'go to' another number in the programme.

A very simple programme in BASIC illustrating the principle, with statement numbers on the left hand side, could be:

1 Think of any number between 10 and 100 (statement)
2 Divide this number by 8 (statement)
3 Enter the answer, whole number (input)
4 Enter the answer remainder (input)
5 What is the whole number answer? (question)
6 Go to 3
7 What is the remainder?
8 Go to 4.

In practice it would be best to leave spare sequential capacity for amending the programme if and when necessary. Thus a better programme would be:

10 Think of any number between 10 and 100
20 Divide this number by 8
30 Enter the answer, whole number
40 Enter the answer, remainder
50 What is the whole number answer?
60 Go to 30
70 What is the remainder?
80 Go to 40.

This leaves scope for amendment. Say you might forget the number you thought of. There is no record of this in the computer memory, only the original number divided by 8 as an answer in 30 and 40. So an additional entry is made after 10, i.e.

11 Enter this number.

At a later stage, a facility to recall the number is entered, e.g.

41 I have forgotten the original number. What was it?
42 Go to 11.

The widespread adoption of BASIC (with its various different dialects) for home, office and small business computers, whereby the computer can be instructed in words, does to a large extent simplify understanding of computer programming. Successful programming, however, still demands skill in order to programme in the most efficient way, and can also be time consuming. The majority of people using computers of this type are generally happier working with bought programmes in casettes (software), or copying from listings (virtually 'skeleton' programmes), rather than composing their own programmes.

Other computer languages include FORTRAN (particularly

favoured for mathematical work); COBOL (particularly favoured for large business data processing as well as being extremely effective for audio-visual tuition); PASCAL (a fairly recent addition implementing the best ideas in computer languages over the last twenty years); ASSEMBLER: ALGOL: PL/I: FORTH: APL; and SORT (assembly language). In addition there are various word-processing systems.

Index

Index

RADIO CONTROL FOR MODELLERS

R. H. WARRING

Radio-controlled models are the modern hobby interest. You can fly your own aeroplane, drive your own powerboat, helm a racing yacht, or race or rally cars at minimal cost and at no personal risk! And modern radio-control sets for models are easier to operate than a domestic radio. They do not need tuning or other adjustments. There is no 'wiring up' to do. The various units which go to make up a complete installation are prewired, ready to plug together, switch on – and go! But it is not quite as simple as that. There is an almost bewildering selection of radio-controlled equipment available, at prices ranging from under £20 for a 'Combo' up to over £400 for a sophisticated 'complete outfit'. You can easily spend a lot more on equipment than you need to do; or equally find that 'minimum cost' equipment has no chance of providing you with what you really require.

This book gives you all the answers you need to select, install and operate radio-controlled models and achieve satisfying, rewarding results. It shows how to select the right type of equipment for the type of model you have in mind (not buying more, or less, performance than you actually need)...how to install your radio equipment in model aircraft, boats, cars (requirements are different)...and how to operate radio-controlled models successfully. There is information on inexpensive ways of operating additional controls...even a chapter on aerial photography using radio-controlled models.

A thoroughly practical book throughout, written in non-technical language by an author with vast experience in R/C models and other practical hobby-interests, and well-known writer on these subjects. It provides a wealth of information and advice which should ensure that the reader gets the maximum satisfaction and enjoyment from one of today's most enthralling hobbies.

Illustrated with 30 photographs and 49 line drawings.

ISBN 0-7188-2519-5

'The book will certainly benefit newcomers to the world of radio control modelling and answer most of the questions originating from such a source'.

MODEL BOATS

INTEGRATED CIRCUITS:
How to make them work

R. H. WARRING

Integrated circuits – or ICs – are largely replacing transistors in all forms of electronic equipment for the home and industry. The modern electronics engineer automatically adopts them as standard practice. This book offers a completely practical introduction for the amateur to the fascinating world of using ICs, in the home or workshop, and turning them into working circuits.

ICs are 'complete' or near-complete circuits which normally need only a few external components added to produce a working electronic device. They are extremely compact and efficient in performance. Due to large-scale manufacture, they are also relatively inexpensive. The problem for the amateur is knowing which IC is suitable for his particular needs – and then how to incorporate it into a working circuit. The author answers these problems by describing the different 'families' of ICs, how to identify their connections, and how they are connected to external components to make working circuits, usually with only the addition of a few resistors and a capacitor or two. One of the great advantages of using ICs, in fact, is the small total number of components usually required and their lower total cost (including the IC) compared with building a similar circuit from separate components.

There are literally thousands of different ICs available today, from simple 'Op-Amps' to complete digital circuits. The author has selected representative types, all of which are readily available, in describing and illustrating eighty-four working circuits. These range from simple voltage regulators to complete radios and electronic organs. Many other useful projects are also included, like a car thief alarm; ice warning indicator; filters and Hi-Fi tone controls; pulse generators; infra-red transmitter and receiver; electronic rev counter; quartz crystal clock; and many, many more. It will prove, therefore, a vital book for anyone interested in, or in any way concerned with, modern electronics practice.

Includes 84 working circuits.

ISBN 0-7188-2343-5

AMATEUR RADIO

GORDON STOKES & PETER BUBB

Amateur radio knows practically no frontiers. Providing conditions permit, it is now possible to 'broadcast' almost anywhere in the world. Use of artificial satellites designated for amateur use and the great diversity of modern equipment make exchanges over vast distances speedy and reliable. Foreign language problems can be eliminated through the use of Morse (CW) transmissions. Alternatively, communication may take place orally, or with fellow enthusiasts as close as the next village, street or town.

This book aims to provide all the relevant background information to students for the Radio Amateur's Examination, leading to the Amateur Radio Licence issued by the Home Office. It describes in detail what radio is and how it works, including essential elementary mathematical formulae to explain first principles. All the basic aspects of transistors, modulation, receivers, transmitters, antennas, transmission lines and HF and VHF signals are dealt with in turn. The important subject of measurement, both to ensure that operation is within the power limitations set by the Home Office and that radiation is within the permitted amateur bands, is fully covered. Portable stations operating at less than 5 watts and video transmissions for amateurs are also described.

The text is comprehensively illustrated with line drawings and photographs. Peter Bubb is an established lecturer and coach for the Radio Amateurs' Examination and Gordon Stokes is an established author who operates his own amateur station. As the authors rightly say, amateur radio is considered by those who practise it to be the most rewarding of all hobbies.

Illustrated with 11 photographs and 74 line drawings.

ISBN 0-7188-2477-6

MODERN TRANSISTOR RADIOS
('See and Make' series)

R. H. WARRING

The circuits for nine different transistor radios are described and illustrated in this book. Each is simple to make, and should give good listening results through headphones. Most can also be adapted to work through loudspeakers.

ISBN 0-7188-2160-2

UNDERSTANDING ELECTRONICS

R. H. WARRING

This is a basic book for anyone who wants – or needs – to gain an understanding of the components and circuitry which combine to make up this modern field of technical expertise. It is of particular value to anyone embarking on electronics for the first time at hobby level, or for the would-be practitioner who is still in need of some grounding in the elements of the subject.

Each chapter is clearly set out and deals with a particular aspect of electronics. The accompanying definitions, simple mathematical equations and formulae, and the explicit line drawings make each stage fully intelligible to the layman. It also offers him the opportunity to build up and experiment with transistorized circuitry in 'breadboard' or printed-circuit form.

The big problem that faces most beginners is interpreting circuit diagrams in order to make an accurate working circuit. The author describes and illustrates fully the meaning of symbols, layouts and various methods of construction, including tagboard assembly, breadboards and printed circuits. There are also helpful practical tips on circuit assembly.

Careful study of this book will ensure a comprehensive understanding of how modern electronic components operate and also the working of radio, television, etc., and the function of transformers and batteries. It will also provide the groundwork which will make possible the understanding of the usually more elaborate circuits described in so many of the books and magazines dealing with modern electronics.

Illustrated with 136 line drawings.

ISBN 0-7188-2327-3

'This outline of a popular-interest subject is sufficiently up to date to include most of the components now available to the home constructor, and it earns a "highly recommended" label.'
THE SCHOOL LIBRARIAN